Solutions Manual

Solutions Manual

to accompany

Lehninger • Nelson • Cox

Principles of Biochemistry

Second Edition

Frederick Wedler

Robert Bernlohr

Ross Hardison

Teh-Hui Kao

Ming Tien

Pennsylvania State University

Worth Publishers

Solutions Manual by Frederick Wedler, Robert Bernlohr, Ross Hardison, Teh-Hui Kao, and Ming Tien

to accompany

Lehninger • Nelson • Cox **Principles of Biochemistry**, Second Edition

Printed in the United States of America
ISBN: 0-87901-729-5
Printing: 1   2   3   4   5
Year: 98   97   96   95   94

Worth Publishers
33 Irving Place
New York, NY 10003

Contents

Preface

Each chapter-end question from the Lehninger • Nelson • Cox **Principles of Biochemistry** text is reprinted in this manual, and its detailed solution immediately follows the question. Great care has been taken to ensure that the solutions are correct, complete, and informative. Throughout this manual, the final answer to numerical problems has been rounded off to reflect the number of significant figures in the data. Due to this rigorous and consistent use of significant figures, some of the answers differ from those given in the Appendix of the text.

In addition to its obvious usefulness as a source of solutions for checking homework assignments, the manual can also help you to decide which problems to assign and which to present in class. You may reproduce and post solutions you assign.

These solutions are available on disk for use on IBM PC (and true compatibles) or Macintosh microcomputers to instructors who adopt the textbook. Please contact Worth Publishers for more information.

CHAPTER **2**   **Cells**

1. *The Size of Cells and Their Components* Given their approximate diameters, calculate the approximate number of (a) hepatocytes (diameter 20 μm), (b) mitochondria (diameter 1.5 μm), and (c) actin molecules (diameter 3.6 nm) that can be placed in a single layer on the head of a pin (diameter 0.5 mm). Assume each structure is spherical. The area of a circle is πr^2, where π = 3.14.

> *Answer*
> Radius of pinhead = 0.5 mm/2 = 0.25mm
> Area of pinhead   = (3.14) (0.25 mm)^2 = 3.14(0.0625) mm^2
>   = 0.196 mm^2
>
> (a) Hepatocytes:
>   Radius = 20 μm/2 = 10 μm = 0.01 mm
>   Cross-sectional area = 3.14(0.01 mm)^2 = 3.14 x 10^{-4} mm^2
>   The number of hepatocytes that could fit on a pinhead is obtained by dividing the pinhead area by the cross-sectional area of one hepatocyte:
>   0.196 mm^2 / (3.14 x 10^{-4} mm^2) = 625 $\approx$ 600 hepatocytes[*]
>
> (b) Mitochondria:
>   Radius = 1.5 μm/2 = 0.75 μm = 7.5 x 10^{-4} mm
>   Cross-sectional area = 3.14(7.5 x 10^{-4} mm)^2 = 3.14(56.3 x 10^{-8}) mm^2
>   = 1.8 x 10^{-6} mm^2
>   The number of mitochondria that could fit on a pinhead is
>   0.196 mm^2 / 1.8 x 10^{-6} mm^2 = 1.1 x 10^5 $\approx$ 1 x 10^5 mitochondria
>
> (c) Actin molecules:
>   Radius = 3.6 nm/2 = 1.8 nm = 1.8 x 10^{-9} m = 1.8 x 10^{-6} mm
>   Cross-sectional area = 3.14(1.8 x 10^{-6} mm)^2 = 3.14(3.24 x 10^{-12}) mm^2
>   = 1.02 x 10^{-11} mm^2
>   The number of actin molecules that could fit on a pinhead is
>   0.196 mm^2 / (1.02 x 10^{-11} mm^2) = 1.92 x 10^{10} $\approx$ 2 x 10^{10} actin molecules

[*]*Significant figures.* In multiplication and division, the answer can be expressed with no more significant figures than the least precise value in the calculation. Since some of the data in these problems are derived from measured values, we must round off the calculated answer to reflect this. In this case, the cell radius (10 μm) has only one significant figure, so the final answer (600 cells) can be expressed with no more than one significant figure. It will be standard practice in these expanded answers to first show the calculated answer, then round it off to the proper number of significant figures. This may lead to some slight differences from answers given in Appendix B of the text.

2. *Number of Solute Molecules in the Smallest Known Cells*   Mycoplasmas are the smallest
    known cells. They are spherical and have a diameter of about 0.33 μm. Because of their small
    size they readily pass through filters designed to trap larger bacteria. One species, *Mycoplasma
    pneumoniae*, is the causative organism of the disease primary atypical pneumonia.

    (a) D-Glucose is the major energy-yielding nutrient of mycoplasma cells. Its concentration
         within such cells is about 1.0 mM. Calculate the number of glucose molecules in a single
         mycoplasma cell. Avogadro's number, the number of molecules in 1 mol of a nonionized
         substance, is 6.02 x 10^{23}. The volume of a sphere is 4/3 πr^3.

    (b) The first enzyme required for the energy-yielding metabolism of glucose is hexokinase (M_r
         100,000). Given that the intracellular fluid of mycoplasma cells contains 10 g of hexokinase
         per liter, calculate the molar concentration of hexokinase.

         Answer
         (a) The radius of a mycoplasma cell is
              0.33 μm/2 = 0.167 μm = 1.67 x 10^5 cm
              From this, the volume of the cell can be calculated as
              (4/3)(3.14)(1.67 x 10^5 m)^3 = 1.95 x 10^{14} cm^3 or mL
              The concentration of glucose is
              1 mM = 1 mmol/L = 10^3 mmol/mL = 10^6 mol/mL
              Multiplying the concentration in mol/mL by Avogadro's number gives
              10^6 mol/mL(6.02 x 10^{23} molecules/mol) = 6.02 x 10^{17} molecules/mL
              Multiplying the volume of the mycoplasma (in mL) by glucose concentration (in
              molecules/mL) gives
              (1.95 x 10^{14} mL)(6.02 x 10^{17} molecules/mL) = 1.1 x 10^4 glucose molecules per
              mycoplasma cell

         (b) Molecular weight can be expressed in units of g/mol, so the intracellular concentration
              of hexokinase can be calculated as
              (10 g/L)/(10^5 g/mol) = 10^4 mol/L = 10^4 M

3. *Components of* **E. coli**   *E. coli* cells are rod-shaped, about 2 μm long and 0.8 μm in diameter.
    The volume of a cylinder is $\pi r^2 h$, where h is the height of the cylinder.

    (a) If the average density of *E. coli* (mostly water) is 1.1 x 10^3 g/L, what is the weight of a
         single cell?

    (b) The protective cell wall of *E. coli* is 10 nm thick. What percentage of the total volume of
         the bacterium does the wall occupy?

    (c) *E. coli* is capable of growing and multiplying rapidly because of the inclusion of some
         15,000 spherical ribosomes (diameter 18 nm) in each cell, which carry out protein synthesis.
         What percentage of the total cell volume do the ribosomes occupy?

         Answer
         (a)   The volume of a single *E. coli* cell can be calculated from $\pi r^2 h$:
                π(0.4)^2(2) = 3.14(4 x 10^5 cm)^2(2 x 10^4 cm) = 1.0 x 10^{12} cm^3 = 1 x 10^{15} L
                Density (g/L) multiplied by volume (L) gives the weight of a single cell:
                (1.1 x 10^3 g/L)(10^{15} L) = 1.1 x 10^{12} g $\approx$ 1 pg

(b)   Wall thickness = 10 nm = 0.01 μm, compared with the cell radius of 0.40 μm. The volume inside the cell wall therefore has a radius of (0.40 - 0.01) μm = 0.39 μm. The difference in volume between these two cylinders, compared with the total cell size, expressed as a percentage, is the percentage occupied by cell wall:

$$\frac{(100)\pi(0.40)^2(2.00) \ \mu m^3 - \pi(0.39)^2(1.98) \ \mu m^3}{\pi(0.40)^2(2.00) \ \mu m^3} = 5.9\% \approx 6\%$$

(c)   Ribosomal radius = 9 nm = 0.9 x 10^{-6} cm
Volume of a single ribosome = 4/3 πr^3 = 1.33(3.14)(0.9 x 10^{-6} cm)^3
= 3.0 x 10^{-18} cm^3
This volume, multipled by the total number of ribosomes, gives the total volume occupied by ribosomes:
15,000(3.0 x 10^{-18} cm^3) = 4.6 x 10^{-14} cm^3 = 0.046 x 10^{-12} cm^3
Given a cell volume of 1.0 x 10^{-12} cm^3, ribosomes constitute about 4.6% of the total *E. coli* cell volume.

4.   *Genetic Information in* **E Coli** *DNA*   The genetic information contained in DNA consists of a linear sequence of successive code words, known as codons. Each codon is a specific sequence of three nucleotides (three nucleotide pairs in double-stranded DNA), and each codon codes for a single amino acid unit in a protein. The molecular weight of an *E. coli* DNA molecule is about 2.5 x 10^9. The average molecular weight of a nucleotide pair is 660, and each nucleotide pair contributes 0.34 nm to the length of DNA.

(a)   Calculate the length of an *E. coli* DNA molecule. Compare the length of the DNA molecule with the actual cell dimensions. How does the DNA molecule fit into the cell?

(b)   Assume that the average protein in *E. coli* consists of a chain of 400 amino acids. What is the maximum number of proteins that can be coded by an *E. coli* DNA molecule?

Answer

(a)   The number of nucleotide pairs in the DNA molecule can be calculated by dividing the molecular weight of DNA by that of a single pair:
(2.5 x 10^9)/(0.66 x 10^3) = 3.8 x 10^6 pairs
Multiplying the number of pairs by the length per pair gives
(3.8 x 10^6 pairs) (0.34 nm/pair) = 1.28 x 10^6 nm ≈ 1.3 mm

The cell dimension is 2.0 μm (from Problem 3), or 0.002 mm, which means that the DNA is 1.3 mm/0.002 mm or 650 times longer than the cell. The DNA must be tightly coiled to fit into the cell.

(b)   Since the DNA molecule has 3.8 x 10^6 nucleotide pairs, as calculated in (a), it must have 1/3 this number triplet codons:
(3.8 x 10^6)/3 = 1.26 x 10^6 codons
If each protein has an average of 400 amino acids, each requiring one codon, the number of proteins that can be coded by *E. coli* DNA is:
(1.26 x 10^6)/400 = 3,156 ≈ 3,200 proteins per cell

5.   *The High Rate of Bacterial Metabolism*   Bacterial cells have a much higher rate of metabolism than animal cells. Under ideal conditions some bacteria will double in size and divide in 20 min, whereas most animal cells require 24 h. The high rate of bacterial metabolism requires a high ratio of surface area to cell volume.

(a) Why would the surface-to-volume ratio have an effect on the maximum rate of metabolism?

(b) Calculate the surface-to-volume ratio for the spherical bacterium, *Neisseria gonorrhoeae* (diameter 0.5 μm), responsible for the disease gonorrhea. Compare it with the surface-to-volume ratio for globular amoeba, a large eukaryotic cell of diameter 150 μm. The surface area of a sphere is $4\pi r^2$.

Answer

(a)   Metabolic rate is limited by diffusion, which in turn is limited by the surface area of the cell. As the ratio of surface area to volume decreases, the rate of diffusion cannot keep up with the rate of metabolism within the cell.

(b)   For a sphere, surface area $= 4\pi r^2$ and volume $= 4/3 \, \pi r^3$. The ratio of these two is the surface-to-volume ratio, S/V, which equals $3/r$ or $6/D$, where D = diameter. Thus, rather than calculating S and V separately for each cell, one can rapidly calculate and compare S/V ratios for cells of different diameters.

S/V for *N. gonorrhoeae* is $6/(0.5 \, \mu m) = 12 \, \mu m^{-1}$

S/V for globular amoeba $= 6/(150) \, \mu m = 0.04 \, \mu m^{-1}$

$$\frac{S/V \text{ for bacteriums}}{S/V \text{ for amoeba}} \qquad \frac{12 \, \mu m^{-1}}{0.04 \, \mu m^{-1}} \qquad \approx 300$$

6.   *A Strategy to Increase the Surface Area of Cells*   Certain cells whose function is to absorb nutrients, e.g., the cells lining the small intestine or the root hair cells of a plant, are optimally adapted to their role because their exposed surface area is increased by microvilli. Consider a spherical epithelial cell (diameter 20 μm) lining the small intestine. Since only a part of the cell surface faces the interior of the intestine, assume that a "patch" corresponding to 25% of the cell area is covered with microvilli. Furthermore, assume that the microvilli are cylinders 0.1μm in diameter, 1.0 μm long, and spaced in a regular grid 0.2 μm on center.

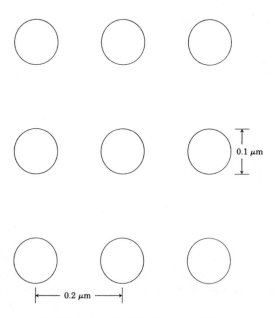

Arrangement of microvilli on the "patch"

(a) Calculate the number of microvilli on the patch.
(b) Calculate the surface area of the patch, assuming it has no microvilli.
(c) Calculate the surface area of the patch, assuming it does have microvilli.
(d) What percentage improvement of the absorptive capacity (reflected by the surface-to-volume ratio) does the presence of microvilli provide?

Answer

(a)   The total surface area of an epithelial cell is
$4\pi r^2 = 4(3.14)(10 \ \mu m)^2 = 1256 \ \mu m^2$
25% of this is exposed, or 314 μm^2 Taking the square-root of this number, the patch can be viewed as a square that is 17.7 μm on a side. The number of microvilli on a side can be calculated by dividing the side dimension by the spacing:
17.7 μm / 0.2 μm = 88.6.
Squaring this value gives the total number of microvilli $\approx$ 7,900.

(b)   The surface area calculated in (a) $= 314 \ \mu m^2$
$= 3.14 \times 10^2 (10^{-6} \ m)^2$
$= 3.14 \times 10^{-10} \ m^2 \ \approx \ 3.1 \times 10^{-10} \ m^2$

(c)   The surface area of one microvillus is $(\pi r^2 + \pi Dh)$, where D = diameter of a cylinder:
$(3.14)(0.05 \ \mu m)^2 \ + \ (3.14)(0.1)(1) \ \mu m^2 \ \approx \ 0.314 \ \mu m^2$
The area occupied by 7850 microvilli = $7850(0.314)\mu m^2 = 2465 \ \mu m^2$.
The total area of the patch is the area covered by microvilli *plus* the area not covered by microvilli. The area of the patch covered by microvilli is 7850 x (surface area of end of microvillus).
Area covered $= 7850(3.14)(0.05)^2 \ \mu m^2 \ = \ 61.6 \ \mu m^2$
Area not covered $= 3.14 \times 10^{-10} \ m^2 - 61.6 \ \mu m^2 \ = \ (314 - 62) \ \mu m^2$
$= 252 \ \mu m^2$
Total area of microvilli-covered patch = $(2465 + 252) \ \mu m^2 = 2717 \ \mu m^2$
$\approx 3 \times 10^{-9} \ m^2$

(d)   The surface area of the patch without microvilli = 314 μm^2, and the surface area with microvilli = 2717 μm^2. Therefore, the improvement in surface area, expressed as a percentage, is
$$\frac{(2717 - 314) \ \mu m^2 \ (100)}{314 \ \mu m^2} = 765\% \ \approx \ 770\%$$

7. **Fast Axonal Transport**   Some neurons have long, thin extensions (axons) as long as 2 m. Small membrane vesicles carrying materials essential to axonal function move along microtubules from the cell body to the tip of the axon by kinesin-dependent "fast axonal transport." If the average velocity of a vesicle is 1 $\mu m/s$, how long does it take a vesicle to move the 2 m from cell body to axonal tip? What are the possible advantages of this ATP-dependent process over simple diffusion to move materials to the axonal tip?

Answer   The time it takes for a vesicle moving at 1 μm/s (or 1 x 10^{-6} m/s) to move a
total of 2 m is

$$\frac{2.0 \text{ m}}{1 \times 10^{-6} \text{ m/s}} = 2 \times 10^6 \text{s}$$

$$= \frac{2 \times 10^6 \text{ s}}{(60 \text{ s/min})(60 \text{ min/h}) (24 \text{ h/day})} \approx 23 \text{ days}$$

A primary advantage of an ATP-dependent process is unidirectional motion: the
vesicle is prevented from randomly diffusing in three dimensions, which would take a
much longer time to accomplish the same effect, especially for a high-molecular-
weight vesicle. Furthermore, since the process is energy-requiring, it can be
regulated by the levels of ATP available and in response to other neurochemical
signals.

8. *Toxic Effects of Phalloidin* Phalloidin is a toxin produced by the mushroom *Amanita
    phalloides*. It binds specifically to actin microfilaments and blocks their disassembly.
    Cytochalasin B is another toxin, which blocks microfilament assembly from actin monomers (see
    p. 42 of text).

    (a) Predict the effect of phalloidin on cytokinesis, phagocytosis, and amoeboid movement, given
         the effects of cytochalasins on these processes.

    (b) A specific antibody (a protein of M_r = 150,000) binds actin tightly and is found to block
         microfilament assembly in vitro (in the test tube). Would you expect this antibody to mimic
         the effects of cytochalasin in vivo (in living cells)?

    Answer
    (a) Cytokinesis, phagocytosis, and amoeboid movement all depend on microfilament
         disassembly and assembly, so phalloidin would have a blocking action on these
         processes.

    (b) Yes, the antibody binds to areas of the actin monomers that must associate in order to
         form microfilaments.

9. *Osmotic Breakage of Organelles* In the isolation of cytosolic enzymes, cells are often broken in
    the presence of 0.2 M sucrose to prevent osmotic swelling and bursting of the intracellular
    organelles. If the desired enzymes are in the cytosol, why is it necessary to be concerned about
    possible damage to particulate organelles?

    Answer Breakage of lysosomes releases lytic (cleaving) enzymes that break down proteins
    and other polymeric biomolecules. Peroxisomes release enzymes and chemicals that cause
    oxidation, which often either inactivates enzymes or potentiates them for further breakdown.

CHAPTER 3 Biomolecules

1. **Vitamin C: Is the Synthetic Vitamin as Good as the Natural One?** One claim put forth by purveyors of health foods is that vitamins obtained from natural sources are more healthful than those obtained by chemical synthesis. For example, it is claimed that pure L-ascorbic acid (vitamin C) obtained from rose hips is better for you than pure L-ascorbic acid manufactured in a chemical plant. Are the vitamins from the two sources different? Can the body distinguish a vitamin's source?

   Answer   The properties of the vitamin are determined by its chemical structure. Since vitamin molecules from the two sources, in pure form, are structurally identical, their properties are identical, and no organism can distinguish the source. If, however, vitamin preparations contain impurities with different physical and biological properties, then the biological effects of the mixture of ascorbic acid with these impurities may vary with the source. The ascorbic acid in such preparations, however, will be identical.

2. **Identification of Functional Groups** Figure 3-5 shows the common functional groups of biomolecules. Since the properties and biological activities of biomolecules are largely determined by their functional groups, it is important to be able to identify them. In each of the molecules, identify the constituent functional groups.

$$H_2N-\overset{\overset{\displaystyle H}{|}}{\underset{\underset{\displaystyle H}{|}}{C}}-\overset{\overset{\displaystyle H}{|}}{\underset{\underset{\displaystyle H}{|}}{C}}-OH$$

Ethanolamine

(a)

$$\begin{array}{c} H-\overset{\overset{\displaystyle H}{|}}{C}-OH \\ H-\overset{}{C}-OH \\ H-\overset{}{C}-OH \\ \underset{H}{|} \end{array}$$

Glycerol

(b)

$$HO-\overset{\overset{\displaystyle O}{\|}}{P}-O^-$$

$$\overset{\overset{\displaystyle H}{\diagdown}}{\underset{H\diagup}{}}C=\overset{\overset{\displaystyle O}{|}}{C}-COOH$$

Phosphoenolpyruvic acid, an intermediate in glucose metabolism

(c)

$$\begin{array}{c} COOH \\ H_2N-\overset{}{C}-H \\ H-\overset{}{C}-OH \\ \underset{}{C}H_3 \end{array}$$

Threonine, an amino acid

(d)

$$\begin{array}{c} HO\diagdown_{\displaystyle C}\diagup^{\displaystyle O} \\ CH_2 \\ CH_2 \\ NH \\ C=O \\ H-C-OH \\ CH_3-C-CH_3 \\ CH_2OH \end{array}$$

Pantothenic acid, a vitamin

(e)

$$\begin{array}{c} H\diagdown_{\displaystyle C}\diagup^{\displaystyle O} \\ H-C-NH_2 \\ HO-C-H \\ H-C-OH \\ H-C-OH \\ CH_2OH \end{array}$$

D-Glucosamine

(f)

Answer

(a)   —NH$_2$ = amino; —OH = hydroxyl

(b)   —OHs = hydroxyl (three)

(c)   —PO$_3^-$H = phosphoryl (in its ionized form); —COOH = carboxyl

(d)   —COOH = carboxyl; —NH$_2$ = amino; —OH = hydroxyl; —CH$_3$ = methyl (two)

(e)   —COOH = carboxyl; —CO—NH— = amido; —OH = hydroxyl (two); —CH$_3$ = methyl

(f)   —CHO = aldehyde; —NH$_2$ = amino; —OHs = hydroxyl (four)

3.   ***Drug Activity and Stereochemistry***   The quantitative differences in biological activity between the two enantiomers of a compound are sometimes quite large. For example, the D isomer of the drug isoproterenol, used to treat mild asthma, is 50 to 80 times more effective as a bronchodilator than the L isomer. Identify the chiral center in isoproterenol. Why would the two enantiomers have such radically different bioactivity?

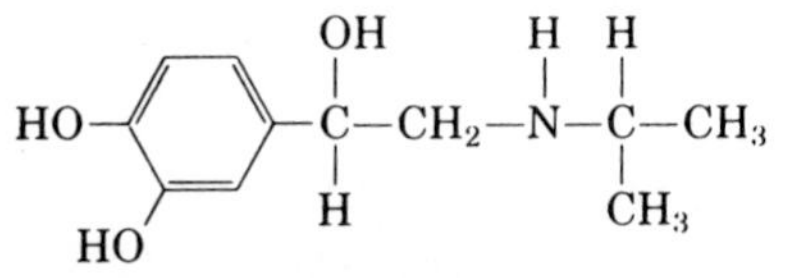

Isoproterenol

Answer   A chiral center occurs at a carbon atom that is bonded to and surrounded by four different groups. One consequence of a single chiral center is that the molecule has two enantiomers, usually designated as D and L or (in the absolute stereochemical convention) *S* and *R*. In isoproterenol, the chiral center must be at the only atom with four different groups around it.

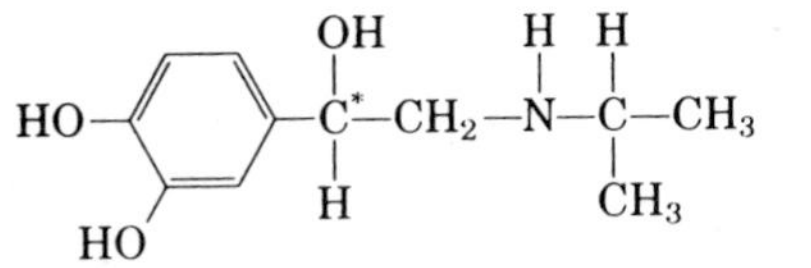

The bioactivity of a drug is the result of interaction with a biological "receptor," a protein molecule with a binding site that is also chiral and stereospecific. As a result, the interaction of the D isomer of a drug with the chiral receptor site will differ from the interaction of the L isomer with that site.

4. ***Drug Action and Shape of Molecules*** Some years ago two drug companies marketed a drug under the trade names Dexedrine and Benzedrine. The structure of the drug is shown below.

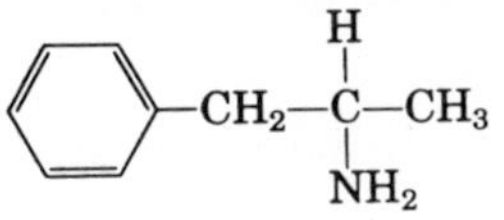

The physical properties (C, H, and N analysis, melting point, solubility, etc.) of Dexedrine and Benzedrine were identical. The recommended oral dosage of Dexedrine (which is still available) was 5 mg/d, but the recommended dosage of Benzedrine was significantly higher. Apparently it required considerably more Benzedrine than Dexedrine to yield the same physiological response. Explain this apparent contradiction.

> ***Answer***   Only one of the two enantionmers of the drug (which has a chiral center) is physiologically active, for reasons described in the answer to Problem 3 (interaction with a stereospecific receptor site). Dexadrine, as manufactured, consists of only the single enantiomer (D-amphetamine) recognized by the receptor site. In contrast, Benzedrine is a racemic mixture (equal amounts of D and L isomers). Thus, a much larger dose of Benzedrine is required to obtain the same effect.

5. ***Components of Complex Biomolecules*** Figure 3-16 shows the structures of the major components of complex biomolecules. For each of the three important biomolecules below (shown in their ionized forms at physiological pH), identify the constituents.

(a) Guanosine triphosphate (GTP), an energy-rich nucleotide that serves as precursor to RNA.

(b) Phosphatidylcholine, a component of many membranes.

(c) Methionine enkephalin, the brain's own opiate.

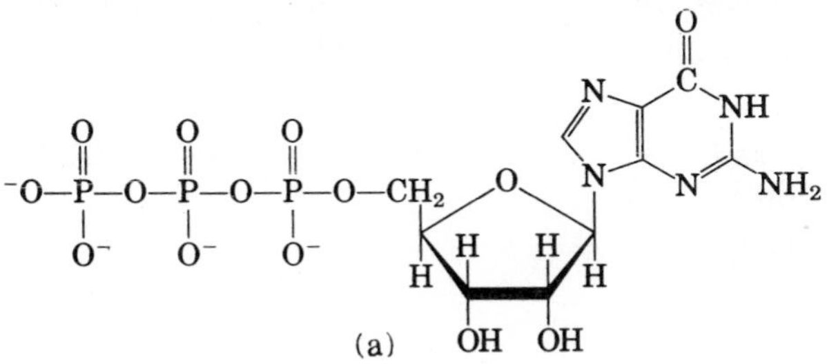

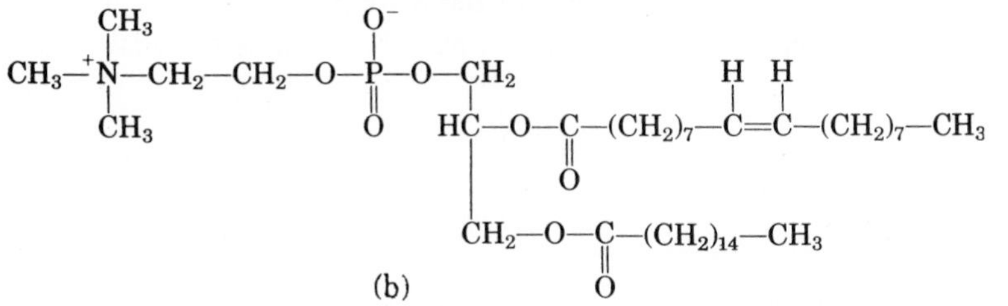

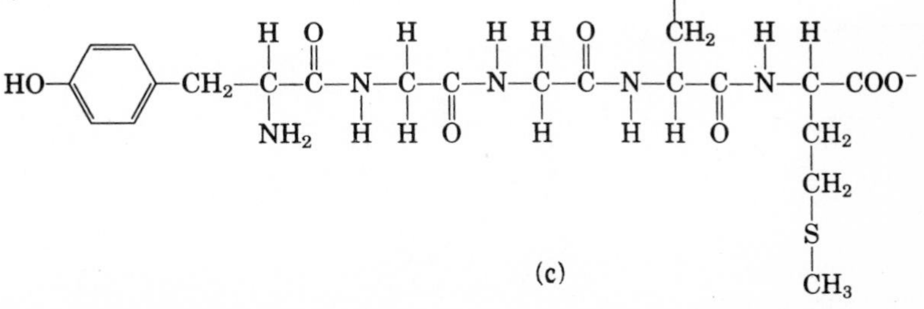

Answer

(a)   Three phosphoric acid molecules (linked by two anhydride bonds), esterified to an α-D-ribose (at the 5′ position), which is attached at C-1 to guanine.

(b)   Choline esterified to a phosphoric acid group, which is esterified to glycerol, which is esterified to two fatty acids, oleic acid and palmitic acid.

(c)   Tyrosine, two glycines, phenylalanine, and methionine, all linked by amide peptide bonds.

6. *Determination of the Structure of a Biomolecule*   An unknown substance, X, was isolated from rabbit muscle. The structure of X was determined from the following observations and experiments. Qualitative analysis showed that X was composed entirely of C, H, and O. A weighed sample of X was completely oxidized, and the amount of H_2O and CO_2 produced was measured. From this quantitative analysis, it was concluded that X contains 40.00% C, 6.71% H, and 53.29% O by weight. The molecular mass of X was determined by a mass spectrometer and found to be 90.00. An infrared spectrum of X showed that it contained one double bond. X dissolved readily in water to give an acidic solution. A solution of X was tested in a polarimeter and demonstrated optical activity.

(a)   Determine the empirical and molecular formula of X.

(b)   Draw the possible structures of X that fit the molecular formula and contain one double bond. Consider *only* linear or branched structures and disregard cyclic structures. Note that oxygen makes very poor bonds to itself.

(c)   What is the structural significance of the observed optical activity? Which structures in (b) does this observation eliminate? Which structures are consistent with the observation?

(d)   What is the structural significance of the observation that a solution of X was acidic? Which structures in (b) are now eliminated? Which structures are consistent with the observation?

(e)   What is the structure of X? Is more than one structure consistent with all the data?

Answer

(a)   From the C, H, and O analyses, and knowing the M_r of X is 90, we can calculate the relative atomic proportions by dividing the % weights by the atomic weights:

Atom	Relative atomic proportion	No. of atoms relative to O
C	[40.00(90)]/[12(100)] = 3	3/3 = 1
H	[6.71(90)]/[1.008(100)] = 6	6/3 = 2
O	[53.29(90)]/[16.0(100)] = 3	3/3 = 1

Thus, the empirical formula is CH_2O, with a formula weight of 12 + 2 + 16 = 30. The molecular formula, based on M_r = 90, must be three times this empirical formula, or $C_3H_6O_3$.

$$
\begin{array}{cccc}
\underset{1}{\text{structure 1}} & \underset{2}{\text{structure 2}} & \underset{3}{\text{structure 3}} & \underset{4}{\text{structure 4}}
\end{array}
$$

$$
\begin{array}{cccc}
\underset{5}{\text{structure 5}} & \underset{6}{\text{structure 6}} & \underset{7}{\text{structure 7}} & \underset{8}{\text{structure 8}}
\end{array}
$$

$$
\begin{array}{cccc}
\underset{9}{\text{structure 9}} & \underset{10}{\text{structure 10}} & \underset{11}{\text{structure 11}} & \underset{12}{\text{structure 12}}
\end{array}
$$

(b)   Structures **1** through **5** can be eliminated because they are unstable enol isomers of the corresponding carbonyl derivatives. Structures **9**, **10**, and **12** can also be eliminated on the basis of their instability: they are hydrated carbonyl derivatives (vicinal diols).

(c)   The presence of optical activity eliminates all structures that lack a chiral center (a C atom surrounded by four different groups). Only structures **6** and **8** have chiral centers.

(d)   X contains an acidic functional group, which structure **8** does not. Structure **6** contains a carboxyl group.

(e)   Structure **6** is the substance X. This compound exists in two enantiomeric forms that cannot be distinguished, even from a measured specific rotation value. One could obtain an absolute stereochemical determination by x-ray crystallography.

CHAPTER 4 Water

1. *Artificial Vinegar* One way to make vinegar (*not* the preferred way) is to prepare a solution of acetic acid, the sole acid component of vinegar, at the proper pH (see Fig. 4-9) and add appropriate flavoring agents. Acetic acid (M_r 60) is a liquid at 25 °C with a density of 1.049 g/mL. Calculate the amount (volume) that must be added to distilled water to make 1 L of simulated vinegar (see Table 4-7).

> *Answer* From Fig. 4-9, the pH of vinegar is 3.0; from Table 4-7, the pK_a of acetic acid is 4.7. Using the Henderson-Hasselbalch equation:
>
> $$pH = pK_a + \log \frac{[A^-]}{[HA]}$$
>
> and the fact that dissociation of HA gives equimolar $[H^+] = [A^-]$ (here, HA is CH_3COOH, and A^- is CH_3COO^-), we can write
>
> $$3.00 = 4.76 + \log ([A^-]/[HA])$$
> $$3.00 - 4.76 = -1.76 = \log ([A^-]/[HA])$$
> $$[HA]/[A^-] = 10^{1.76} = 58$$
>
> Since $[H^+] = [A^-] = 10^{-3}$ at pH 3.0, this value can be substituted into the denominator to give $[HA] = 5.8 \times 10^{-2}$ M = 0.058 mol/L
>
> The units for density are g/mL, and molecular weight can be expressed in g/mol. Dividing density by molecular weight gives
>
> $$\frac{1.049 \text{ g/mL}}{60 \text{ g/mol}} = 0.0175 \text{ mol/mL}$$
>
> Dividing this value into 0.058 mol/L gives the volume of acetic acid needed to prepare 1.0 L of a 0.058 M solution:
>
> $$\frac{0.058 \text{ mol/L}}{0.0175 \text{ mol/mL}} = 3.32 \text{ mL per liter of solution}$$

2. *Acidity of Gastric HCl* In a hospital laboratory, a 10.0 mL sample of gastric juice, obtained several hours after a meal, was titrated with 0.1 M NaOH to neutrality; 7.2 mL of NaOH was required. The stomach contained no ingested food or drink, thus assume that no buffers were present. What was the pH of the gastric juice?

> *Answer* Multiplying volume (mL) by molar concentration (mol/L) gives the number of moles in the volume added or present. If x is the concentration of gastric HCl (mol/L):
>
> $$(10 \text{ mL}) x = (7.2 \text{ mL})(0.1 \text{ mol/L})$$
> $$x = 0.072 \text{ M gastric HCl}$$
>
> Since by definition, pH = -log $[H^+]$, and since HCl is a strong acid:
>
> $$pH = -\log (7.2 \times 10^{-2}) = 1.1$$

3. *Measurement of Acetylcholine Levels by pH Changes*   The concentration of acetylcholine, a neurotransmitter, can be determined from the pH changes that accompany its hydrolysis. When incubated with a catalytic amount of the enzyme acetylcholinesterase, acetylcholine is quantitatively converted into choline and acetic acid, which dissociates to yield acetate and a hydrogen ion (see structures below). In a typical analysis, 15 mL of an aqueous solution containing an unknown amount of acetylcholine had a pH of 7.65. When incubated with acetylcholinesterase, the pH of the solution decreased to a final value of 6.87. Assuming that there was no buffer in the assay mixture, determine the number of moles of acetylcholine in the 15 mL of unknown.

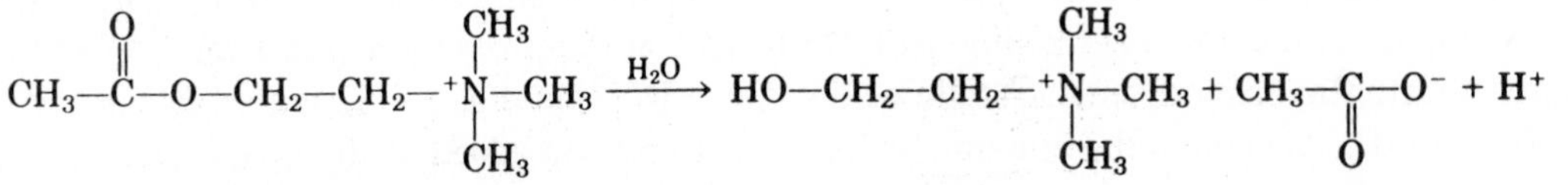

$$CH_3-\overset{O}{\overset{\|}{C}}-O-CH_2-CH_2-\overset{CH_3}{\overset{|}{\underset{\underset{CH_3}{|}}{{}^+N}}}-CH_3 \xrightarrow{H_2O} HO-CH_2-CH_2-\overset{CH_3}{\overset{|}{\underset{\underset{CH_3}{|}}{{}^+N}}}-CH_3 + CH_3-\overset{}{\underset{\underset{O}{\|}}{C}}-O^- + H^+$$

Acetylcholine   Choline   Acetate

Answer Since pH = -log [H$^+$], we can calculate [H$^+$] at the beginning and at the end of the reaction.

At pH 7.65: log [H$^+$] = -7.65   [H$^+$] = $10^{-7.65}$ = 2.24 x 10^{-8} M

At pH 6.87, log [H$^+$] = -6.87   [H$^+$] = $10^{-6.87}$ = 1.35 x 10^{-7} M

The difference in [H$^+$] is

(1.35 - 0.22) x 10^{-7} M = 1.13 x 10^{-7} M

For a volume of 15 mL, or 0.015 L, multiplying volume by molarity gives

0.015 L (1.13 x 10^{-7} mol/L) = 1.7 x 10^{-9} mol of acetylcholine

4. *Significance of the pK$_a$ of an Acid*   One common description of the pK$_a$ of an acid is that it represents the pH at which the acid is half ionized, that is, the pH at which it exists as a 50:50 mixture of the acid and the conjugate base. Demonstrate this relationship for an acid HA, starting from the equilibrium-constant expression.

Answer For the equilibrium   HA $\rightleftharpoons$ H$^+$ + A$^-$, the Henderson-Hasselbalch equation is pH = pK$_a$ + log $\dfrac{[A^-]}{[HA]}$

When the acid is half-ionized, [HA] = [A$^-$], and the ratio [A$^-$]/[HA] = 1.0.

Since log 1.0 = 0, when [HA] = [A$^-$] the log ([A$^-$]/[HA]) term becomes zero and the Henderson-Hasselbalch equation simplifies to pH = pK$_a$.

5. *Properties of a Buffer* The amino acid glycine is often used as the main ingredient of a buffer in biochemical experiments. The amino group of glycine, which has a pK_a of 9.6, can exist either in the protonated form ($-NH_3^+$) or as the free base ($-NH_2$) because of the reversible equilibrium

$$R-NH_3^+ \rightleftharpoons R-NH_2 + H^+$$

(a) In what pH range can glycine be used as an effective buffer due to its amino group?

(b) In a 0.1 M solution of glycine at pH 9.0, what fraction of glycine has its amino group in the $-NH_3^+$ form?

(c) How much 5 M KOH must be added to 1.0 L of 0.1 M glycine at pH 9.0 to bring its pH to exactly 10.0?

(d) In order to have 99% of the glycine in its $-NH_3^+$ form, what must the numerical relation be between the pH of the solution and the pK_a of the amino group of glycine?

Answer

(a) Glycine buffers well in a zone from about one pH unit above to one pH unit below pH 9.6.

(b) Using the Henderson-Hasselbalch equation

$$pH = pK_a + \log \frac{[A^-]}{[HA]}$$

we can write

$$9.0 = 9.6 + \log \frac{[A^-]}{[HA]}$$

$$\frac{[A^-]}{[HA]} = 10^{-0.6} = 0.25$$

which corresponds to a ratio of 1/4. This indicates that the amino group of glycine is about 4/5 (80%) in the protonated form at pH 9.0.

(c) From (b) we know that the amino group is about 20% deprotonated at pH 9.0. Thus in moving from pH 9.0 to pH 9.6 (at which it is 50% deprotonated) 0.3 (30%) of the glycine is titrated. Next, we can calculate from the Henderson-Hasselbalch equation the percentage protonation at pH 10.0:

$$10.0 = 9.6 + \log \frac{[A^-]}{[HA]}$$

$$\frac{[A^-]}{[HA]} = 10^{0.4} = 2.5 = 5/2$$

This ratio indicates that glycine is 5/7 or 71% deprotonated at pH 10.0, an additional 21% deprononation above that (50%) at the pK_a. Thus the total fractional deprotonation in moving from pH 9.0 to 10.0 is (0.30 + 0.21) = 0.51, which corresponds to
0.51 x 0.1 mol = 0.051 mol of KOH required
For a 5 M KOH solution, this corresponds to 0.01 L, or about 10 mL of KOH solution.

(d) Qualitatively, you can reason that the pH must be at least two pH units above the pK_a in order to satisfy this condition. Quantitatively, this can be expressed as a rearranged form of the Henderson-Hasselbalch equation:
$$pH - pK_a = -2$$

6.   ***The Effect of pH on Solubility***   The strongly polar hydrogen-bonding nature of water makes it
an excellent solvent for ionic (charged) species. By contrast, un-ionized, nonpolar organic
molecules, such as benzene, are relatively insoluble in water. In principle, the aqueous solubility
of all organic acids or bases can be increased by deprotonation or protonation of the molecules,
respectively, to form charged species. For example, the solubility of benzoic acid in water is
low. The addition of sodium bicarbonate raises the pH of the solution and deprotonates the
benzoic acid to form benzoate ion, which is quite soluble in water.

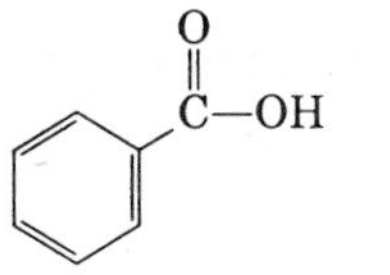

Benzoic acid
$pK_a \approx 5$

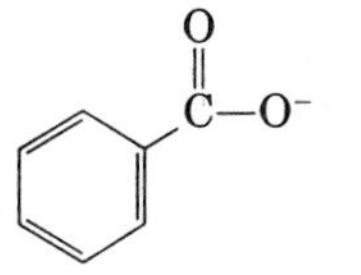

Benzoate ion

Are the molecules in **(a)** to **(c)** more soluble in an aqueous solution of 0.1 M NaOH or 0.1 M
HCl?

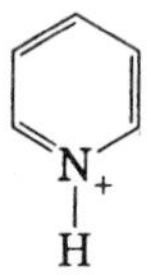

Pyridine ion
$pK_a \approx 5$

(a)

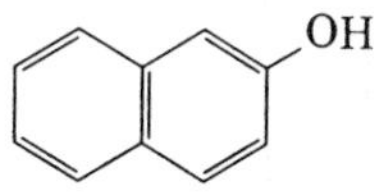

β-Naphthol
$pK_a \approx 10$

(b)

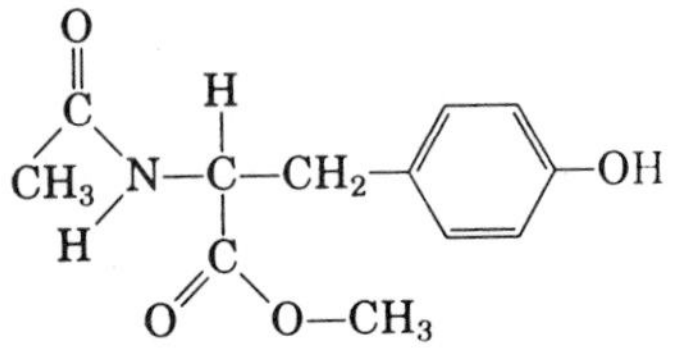

N-Acetyltyrosine methyl ester
$pK_a \approx 10$

(c)

Answer

(a)   Pyridine is ionic in its protonated form and therefore more soluble at low pH, that is,
in 0.1 M HCl.

(b)   β-Naphthol is ionic when *de*protonated and thus more soluble at high pH, that is, in
0.1 M NaOH.

(c)   N-Acetyltyrosine methyl ester is ionic when *de*protonated and thus more soluble in 0.1
M NaOH.

7.   ***Treatment of Poison Ivy Rash***   Catechols substituted with long-chain alkyl groups are the
components of poison ivy and poison oak that produce the characteristic itchy rash (see structure
below). If you were exposed to poison ivy, which of the treatments below would you apply to
the affected area? Justify your choice.

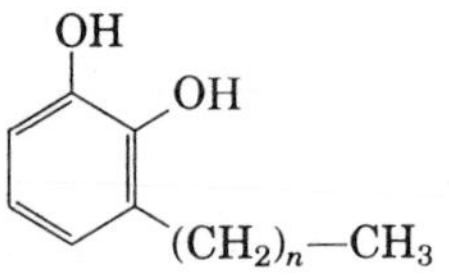

$(CH_2)_n$—CH_3
$pK_a \approx 8$

(a) Wash the area with cold water.

(b) Wash the area with dilute vinegar or lemon juice.

(c) Wash the area with soap and water.

(d) Wash the area with soap, water, and baking soda (sodium bicarbonate).

Answer Soap will help emulsify and dissolve the hydrophobic alkyl groups of the alkylcatechols. The pK_a of catechols is about 8.0 (see text p. 106), and a mildly alkaline solution of bicarbonate ($NaCO_3$) will ionize the catechol $-OH$ group to $-O^-$, making the molecule much more water-soluble. A pure aqueous or acidic solution alone, as in (a) or (b), will not be effective. Thus (d) is the best choice.

8. **pH and Drug Absorption** Aspirin is a weak acid with a pK_a of 3.5 (see structure below). It is absorbed into the blood through the cells lining the stomach and the small intestine. Absorption requires passage through the cell membrane, which is determined by the polarity of the molecule: charged and highly polar molecules pass slowly, whereas neutral hydrophobic ones pass rapidly. The pH of the gastric juice in the stomach is about 1.5 and the pH of the contents of the small intestine is about 6. Is more aspirin absorbed into the bloodstream from the stomach or from the small intestine? Clearly justify your choice.

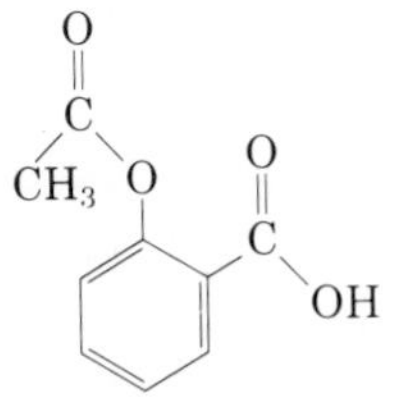

Answer Having a pK_a = 3.5, aspirin exists in its protonated (neutral) form below pH 2.5. At the pH of the intestinal tract, aspirin is in its anionic form. At the pH of the stomach, aspirin is neutral, and absorption will be better in this acidic environment.

9. **Preparation of Standard Buffer for Calibration of a pH Meter** The glass electrode used in commercial pH meters gives an electrical response proportional to the hydrogen-ion concentration. To convert these responses into pH, glass electrodes must be calibrated against standard solutions of known hydrogen-ion concentration. Determine the weight in grams of sodium dihydrogen phosphate ($NaH_2PO_4 \cdot H_2O$; formula weight (FW) 138.01) and disodium hydrogen phosphate (Na_2HPO_4; FW 141.98) needed to prepare 1 L of a standard buffer at pH 7.00 with a total phosphate concentration of 0.100 M (see Table 4-7).

Answer From Table 4-7, the pK_a for the dissociation

$$H_2PO_4^- \rightleftharpoons HPO_4^{2-} + H^+ \text{ is 6.86}$$

Using the Henderson-Hasselbalch equation

$$pH = pK_a + \log\frac{[A^-]}{[HA]}$$

$$7.00 - 6.86 = \log\frac{[A^-]}{[HA]}$$

$$\frac{[A^-]}{[HA]} = 10^{0.14} = 1.38$$

which is quite close to 7/5, that is, 7 parts Na_2HPO_4 to 5 parts $NaH_2PO_4 \cdot H_2O$. Since $[HPO_4^{2-}] + [H_2PO_4^-] = 0.100$ M, $[H_2PO_4^-] = 0.100 - [HPO_4^{2-}]$. Substituting this relationship into

$$\frac{[A^-]}{[HA]} = 1.38 \text{ gives}$$

$$\frac{[H_2PO_4^-]}{0.100 - [HPO_4^{2-}]} = 1.38 \quad \text{Thus we can calculate the concentration of both species.}$$

$$[HPO_4^{2-}] = \frac{0.138}{2.38} = 0.058 \text{ M,}$$

$$[H_2PO_4^-] = 0.100 - [HPO_4^{2-}] = 0.042 \text{ M}$$

The amount needed for 1 L of solution = FW x (mol/L) = FW x M.

For Na_2HPO_4:   (141.98 g/mol)(0.058 M) = 8.23 g/L

For $NaH_2PO_4 \cdot H_2O$:   (138.01 g/mol)(0.042 M) = 5.80 g/L

10. *Control of Blood pH by the Rate of Respiration*

(a) The partial pressure of CO_2 in the lungs can be varied rapidly by the rate and depth of breathing. For example, a common remedy to alleviate hiccups is to increase the concentration of CO_2 in the lungs. This can be achieved by holding one's breath, by very slow and shallow breathing (hypoventilation), or by breathing in and out of a paper bag. Under such conditions, the partial pressure of CO_2 in the air space of the lungs rises above normal. Qualitatively explain the effect of these procedures on the blood pH.

(b) A common practice of competitive short-distance runners is to breathe rapidly and deeply (hyperventilation) for about half a minute to remove CO_2 from their lungs just before running in, say, a 100 m dash. Their blood pH may rise to 7.60. Explain why the blood pH goes up.

(c) During a short-distance run the muscles produce a large amount of lactic acid from their glucose stores. In view of this fact, why might hyperventilation before a dash be useful?

Answer

(a)   Blood pH is controlled by the carbon dioxide-bicarbonate buffer system, as shown in the net equation:

$$CO_2 + H_2O \rightleftharpoons H^+ + HCO_3^-$$

During *hypoventilation* the concentration of CO_2 increases in the lungs and arterial blood, driving the equilibrium to the right and raising the $[H^+]$; that is the pH is lowered.

(b)   During *hyperventilation* the concentration of CO_2 is lowered in the lungs and arterial blood. This drives the equilibrium to the left, which requires the consumption of hydrogen ions. Thus, $[H^+]$ is reduced; that is, the pH is raised from the normal 7.4 value.

(c)   Lactate is a moderately strong acid ($pK_a = 3.86$) that completely dissociates under the physiological conditions:

$$CH_3CHOHCOOH \rightleftharpoons CH_3CHOHCOO^- + H^+$$

This lowers the pH of the blood and muscle tissue. Hyperventilation is useful because it removes hydrogen ions, raising the pH of the blood and tissues in anticipation of the acid build up.

CHAPTER 5   Amino Acids and Peptides

1. **Absolute Configuration of Citrulline**   Is citrulline isolated from watermelons (shown below) a D- or L-amino acid? Explain.

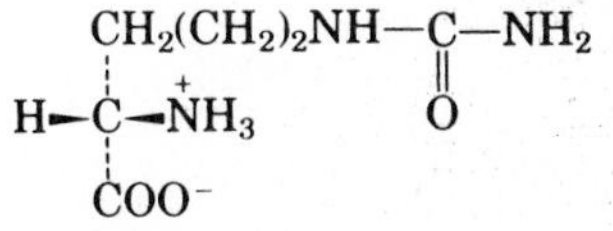

$$CH_2(CH_2)_2NH-\underset{\underset{O}{\|}}{C}-NH_2$$

$$H-\underset{\underset{COO^-}{|}}{\overset{+}{C}}-NH_3$$

Answer   Rotating the molecule by 180° in the plane of the page puts the (most highly oxidized) carboxyl group at the top—analogous to the —CHO group of glyceraldehyde in Figure 5-4—and the amino group on the left (*levo*). Thus the absolute configuration of the citrulline is L.

2. **Relation between the Structures and Chemical Properties of the Amino Acids**   The structures and chemical properties of the amino acids are crucial to understanding how proteins carry out their biological functions. The structures of the side chains of 16 amino acids are given. Name the amino acid that contains each structure and match the R group with the most appropriate description of its properties, (a) to (m). Some of the descriptions may be used more than once.

(a) Small polar R group containing a hydroxyl group; this amino acid is important in the active site of some enzymes.

(b) Provides the least amount of steric hindrance.

(c) R group has $pK_a \approx 10.5$, making it positively charged at physiological pH.

(d) Sulfur-containing R group; neutral at any pH.

(e) Aromatic R group, hydrophobic in nature and neutral at any pH.

(f) Saturated hydrocarbon, important in hydrophobic interactions.

(g) The only amino acid having an ionizing R group with a pK_a near 7; it is an important group in the active site of some enzymes.

(h) The only amino acid having a substituted α-amino group; it influences protein folding by forcing a bend in the chain.

(i) R group has a pK_a near 4 and thus is negatively charged at pH 7.

(j) An aromatic R group capable of forming hydrogen bonds; it has a pK_a near 10.

(k) Forms disulfide cross-links between polypeptide chains; the pK_a of its functional group is about 10.

(l) R group with $pK_a \approx 12$, making it positively charged at physiological pH.

(m) When this polar but uncharged R group is hydrolyzed, the amino acid is converted into another amino acid having a negatively charged R group at pH near 7.

(1) —H   (2) —CH$_3$ (3) 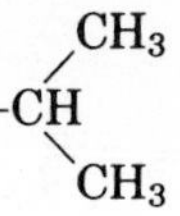 —CH$\Big\langle\begin{array}{l}\text{CH}_3\\\text{CH}_3\end{array}$ (4) 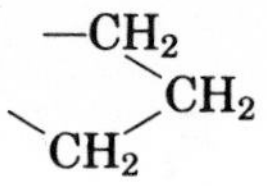 $\begin{array}{c}\text{—CH}_2\\\text{CH}_2\end{array}\Big\rangle\text{CH}_2$

(5) —CH$_2$OH   (6) 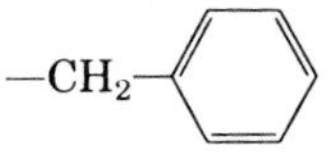 —CH$_2$—⬡   (7) 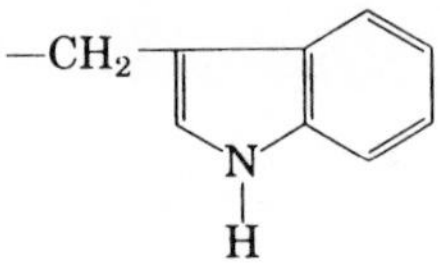 —CH$_2$—[indole]   (8) 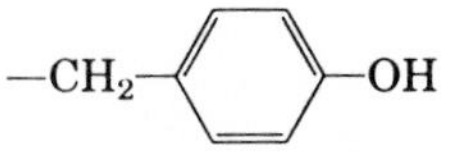 —CH$_2$—⬡—OH

(9) 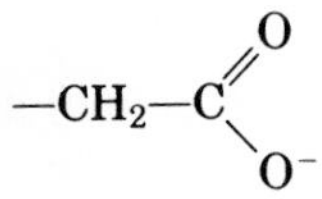 —CH$_2$—C$\begin{array}{c}\text{O}\\\text{O}^-\end{array}$ (10) 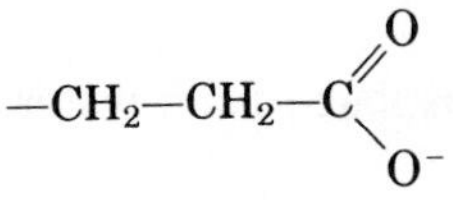 —CH$_2$—CH$_2$—C$\begin{array}{c}\text{O}\\\text{O}^-\end{array}$ (11) —CH$_2$—CH$_2$—S—CH$_3$ (12) —CH$_2$—SH

(13) 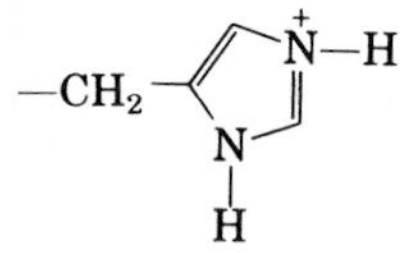 —CH$_2$—[imidazole]   (14) 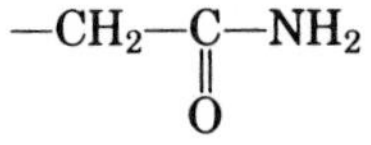 —CH$_2$—C$\begin{array}{c}\\\text{O}\end{array}$—NH$_2$ (15) 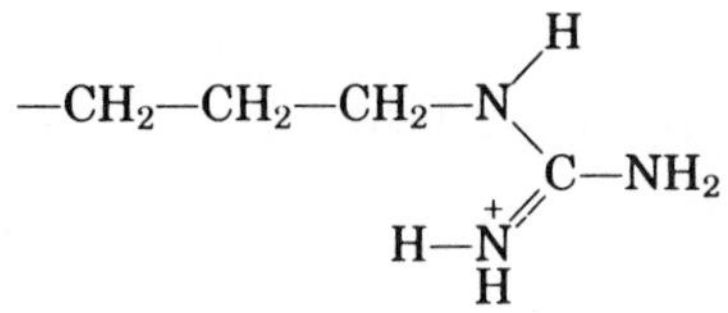 —CH$_2$—CH$_2$—CH$_2$—N$\cdots$

(16) —CH$_2$—CH$_2$—CH$_2$—CH$_2$—$\overset{+}{\text{N}}$H$_3$

Answer

(1) Glycine (b)	**(2)** Analine (f)	**(3)** Valine (f)	**(4)** Proline (h)
(5) Serine (a)	**(6)** Phenylalanine (e)	**(7)** Tryptophan (e)	**(8)** Tyrosine (j)
(9) Aspartate (i)	**(10)** Glutamate (i)	**(11)** Methionine (d)	**(12)** Cysteine (k)
(13) Histidine (g)	**(14)** Asparagine (m)	**(15)** Arginine (l)	**(16)** Lysine (c)

3. *Relationship between the Titration Curve and the Acid-Base Properties of Glycine* A 100 mL solution of 0.1 M glycine at pH 1.72 was titrated with 2 M NaOH solution. During the titration, the pH was monitored and the results were plotted in the graph shown. The key points in the titration are designated I to V on the graph. For each of the statements below, *identify* the appropriate key point in the titration and *justify* your choice.

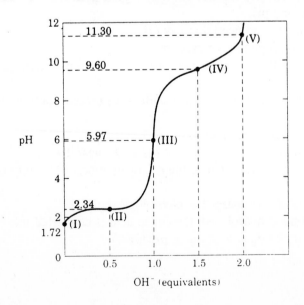

OH$^-$ (equivalents)

Note: Before considering statements (**a**) through (**s**) refer to Figure 5-9. The three species involved in the titration of Gly can be considered in terms of a useful physical analogy. Each ionic species can be viewed as a different floor of a building, each with a different net charge:

$$^+H_3N-CH_2-COOH \qquad\qquad +1$$
$$^+H_3N-CH_2-COO^- \qquad\qquad 0 \text{ (zwitterion)}$$
$$H_2N-CH_2-COO^- \qquad\qquad -1$$

The floors are connected by steep stairways, and each stairway has a landing halfway between the floors. A titration curve traces the path one would follow between the different floors, as the pH changes in response to added OH^-. Recall that the pK_a of an acid represents the pH at which half of the acid is deprotonated. The isoelectric point (pI) is the pH at which the average net charge is zero.

(**a**)   At what point will glycine be present predominantly as $^+H_3N-CH_2-COOH$?
(**b**)   At what point is the *average* net charge of glycine $+\frac{1}{2}$?
(**c**)   At what point is the amino group of half of the molecules ionized?
(**d**)   At what point is the pH equal to the pK_a of the carboxyl group?
(**e**)   At what point is the pH equal to pK_a of the protonated amino group?
(**f**)   At what points does glycine have its maximum buffering capacity?

Answer
(**a**)   I; at the lowest pH—the highest $[H^+]$—the maximum state of protonation occurs.
(**b**)   II; at the first, pK_a (2.34), 0.5 of the protons are removed from the α-carboxyl group (i.e., it is half deprotonated), changing its charge from 0 to -0.5, making the net charge of Gly = (-0.5 + 1) = +0.5.
(**c**)   IV; the α-amino group is half-deprotonated at its pK_a (9.6).
(**d**)   II; from the Henderson-Hasselbalch equation, pH = pK_a + log $([A^-]/[HA])$, when $[A^-]/[HA]$ = 1 or $[A^-]$ = $[HA]$, log(1) = 0 and pH = pK_a.
(**e**)   IV; see answers (c) and (d).
(**f**)   II and IV; in the region of each pK_a, acid donates protons to or base abstracts protons from Gly, with minimal pH changes.

(**g**)   At what point is the *average* net charge zero?
(**h**)   At what point has the carboxyl group been completely titrated (first equivalent point)?
(**i**)   At what point are half of the carboxyl groups ionized?
(**j**)   At what point is glycine completely titrated (second equivalence point)?
(**k**)   At what point is the structure of the predominant species $^+H_3N-CH_2-COO^-$?
(**l**)   At what point do the structures of the predominant species correspond to a 50:50 mixture of $^+H_3N-CH_2-COO^-$ and $H_2N-CH_2-COO^-$?

Answer

(g)   III; pI = (pK_1 + pK_2)/2 = (2.34 + 9.60)/2 = 5.97.

(h)   III; the first equivalence point (for the —COOH, where 1.0 equivalent of OH^- has been added) occurs at pH = 5.97, fully 2.6 pH units away from either pK_a.

(i)   II; see answer (b).

(j)   V; titration is complete at pH 11.3, nearly two pH units above pK_2.

(k)   III; at pI the —COOH group is fully negatively charged, offset by a fully protonated, positively charged —NH_3^+ group.

(l)   IV; this is the second pK_a (9.6).

(m) At what point is the *average* net charge of glycine -1?

(n) At what point do the structures of the predominant species consist of a 50:50 mixture of ^+H_3N—CH_2—COOH and ^+H_3N—CH_2—COO^-?

(o) What point corresponds to the isoelectric point?

(p) At what point is the *average* net charge on glycine -½?

(q) What point represents the end of the titration?

(r) If one wanted to use glycine as an efficient buffer, which points would represent the *worst* pH regions for buffering power?

(s) At what point in the titration is the predominant species H_2N—CH_2—COO^-?

Answer

(m)   V; both groups are fully protonated, producing a neutral amino group and a negatively charged carboxylate.

(n)   II; the —COOH group is half-ionized at pH = pK_1.

(o)   III; see answers to (g) and (k).

(p)   IV; since the net charge is zero at III (the pI) and -1 at V, then at the pK_a between these two points Gly will be half-ionized with a net charge of -½.

(q)   V; Gly is fully titrated once 2.0 equivalents of OH^- have been added.

(r)   I, III, and V; each of these is several pH units removed from either pK_a where the best pH buffering action occurs.

(s)   V; see answers to (m) and (q).

4.   *How Much Alanine Is Present as the Completely Uncharged Species?*   At a pH equal to the isoelectric point, the *net* charge on alanine is zero. Two structures can be drawn that have a net charge of zero (zwitterionic and uncharged forms), but the predominant form of alanine at its pI is zwitterionic.

Zwitterionic   Uncharged

(a) Explain why the form of alanine at its pI is zwitterionic rather than completely uncharged.

(b) Estimate the fraction of alanine present at its pI as the completely uncharged form. Justify your assumptions.

Answer

(a)   The pI of Ala occurs at a pH well above the pK_a for the α-carboxyl group and well below the pK_a for the α-amino group. Hence, both groups will be present predominantly in their charged (ionized) forms.

(b)   The pI of Ala = 6.01, midway between the two pK_a values, 2.34 and 9.69. From the Henderson-Hasselbach equation, pH - pK_a = log ([A⁻]/[HA])

$$\log \frac{[A^-]}{[HA]} = 3.67$$

$$\frac{[A^-]}{[HA]} = 10^{-3.67} = \frac{1}{4678}$$

That is, one molecule in 4678 is still in the form of —COOH. Similarly, at pH = pI one molecule in 4678 is in the form of —NH₂. Thus, the number of molecules with both groups uncharged (—COOH *and* —NH₂) is 1 in (4678)(4678), or 1 in 2.2 x 10⁷.

5.   *Ionization State of Amino Acids*   Each ionizable group of an amino acid can exist in one of two states, charged or neutral. The electric charge on the functional group is determined by the relationship between its pK_a and the pH of the solution, described by the Henderson-Hasselbalch equation.

(a)   Histidine has three ionizable functional groups. Write the relevant equilibrium equations for its three ionizations and assign the proper pK_a for each ionization. Draw the structure of histidine in each ionization state. What is the net charge on the histidine molecule in each ionization state?

Answer

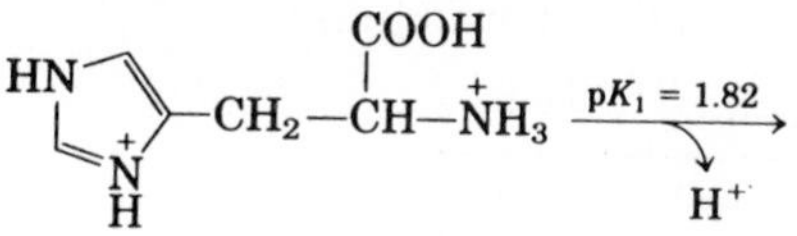

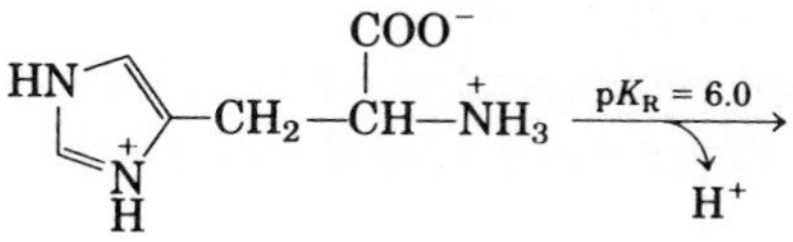

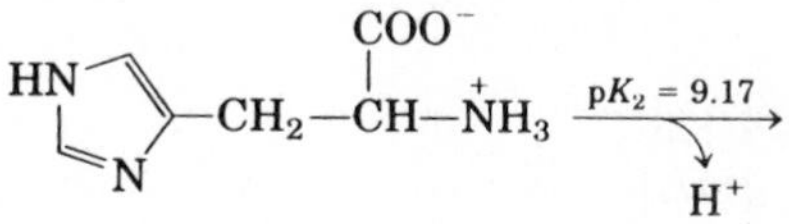

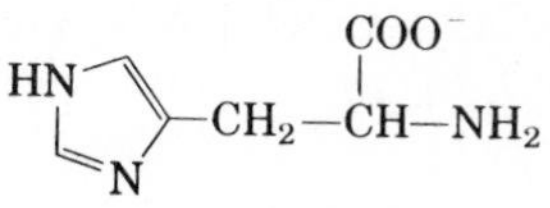

The systematic method used here is as follows: starting with the most highly protonated species (that found at the most acidic pH), find the pK_a for each group in Table 5-1. As base is added, the group with the lowest pK_a will lose its proton first, followed by the group with the next-lowest pK_a, then that with the highest pK_a. (In the following table, R = imidazole.)

	Structure	Total charge
1	$^+H_3N-CH(RH^+)-COOH$	+2
2	$^+H_3N-CH(RH^+)-COO^-$	+1
3	$^+H_3N-CH(R)-COO^-$	0
4	$H_2N-CH(R)-COO^-$	-1

(b) Draw the structures of the predominant ionization state of histidine at pH 1, 4, 8, and 12. Note that the ionization state can be approximated by treating each ionizable group independently.

(c) What is the net charge of histidine at pH 1, 4, 8, and 12? For each pH, will histidine migrate toward the anode (+) or cathode (-) when placed in an electric field?

Answers **(b)** and **(c)** See structures in (a).

pH	Structure	Net charge	Migrates toward:
1	1	+2	Cathode (-)
4	2	+1	Cathode (-)
8	3	0	Does not migrate
12	4	-1	Anode (+)

6. *Preparation of a Glycine Buffer*   Glycine is commonly used as a buffer. Preparation of a 0.1 M glycine buffer starts with 0.1 M solutions of glycine hydrochloride ($HOOC-CH_2-NH_3^+Cl^-$) and glycine ($^-OOC-CH_2-NH_3^+$), two commercially available forms of glycine. What volumes of these two solutions must be mixed to prepare 1 L of 0.1 M glycine buffer having a pH of 3.2? (Hint: See Box 4-2)

Answer We can substitute values for pH and pK_a into the Henderson-Hasselbalch equation to give

$3.2 = 2.34 + \log \dfrac{[A^-]}{[HA]}$

$\dfrac{[A^-]}{[HA]} = 10^{0.86} = 7.24.$

If $[^+H_3N-CH_2-COO^-] = x$, then $[^+H_3N-CH_2-COOH] = 0.1 - x$, so

$\dfrac{x}{0.1 - x} = 7.24$, or

$x = \dfrac{0.724}{8.24} = 0.0879$

Thus

$[^+H_3N-CH_2-COO^-] = 0.0879$ M, by adding 879 mL of 0.1 M Gly

$[^+H_3N-CH_2-COOH] = 0.0121$ M, by adding 121 mL of 0.1 M Gly-HCl

7. *Separation of Amino Acids by Ion-Exchange Chromatography* Mixtures of amino acids are analyzed by first separating the mixture into its components through ion-exchange chromatography. On a cation-exchange resin containing sulfonate groups (see Fig. 5-12), the amino acids flow down the column at different rates because of two factors that retard their movement: (1) ionic attraction between the $-SO_3^-$ residues on the column and positively charged functional groups on the amino acids and (2) hydrophobic interaction between amino acid side chains and the strongly hydrophobic backbone of the polystyrene resin. For each pair of amino acids listed, determine which member will be eluted first from an ion-exchange column by a pH 7.0 buffer.

(a) Asp and Lys
(b) Arg and Met
(c) Glu and Val
(d) Gly and Leu
(e) Ser and Ala

Answer See Table 5-1 for pK_a values for the amino acid side chains. At pH < pI, an amino acid has a positive charge; at pH > pI, it has a negative charge. For any pair of amino acids, the more negatively charged species will pass through the sulfonated resin faster. For two neutral amino acids, the less polar one will pass through more slowly due to its stronger hydrophobic interactions with the polystyrene.

	pI	Net charge (pH 7)	Elution order	Basis for separation
(a) Asp, Lys	2.8, 9.7	-1, +1	Asp, Lys	Charge
(b) Arg, Met	5.7, 10.8	+1, 0	Met, Arg	Charge
(c) Glu, Val	3.2, 6.0	-1, 0	Glu, Val	Charge
(d) Gly, Leu	5.97, 5.98	0, 0	Gly, Leu	Polarity
(e) Ser, Ala	5.68, 6.01	0, 0	Ser, Ala	Polarity

8. *Naming the Stereoisomers of Isoleucine* The structure of the amino acid isoleucine is:

$$
\begin{array}{c}
COO^- \\
| \\
H_3{}^+N-C-H \\
| \\
H-C-CH_3 \\
| \\
CH_2 \\
| \\
CH_3
\end{array}
$$

(a) How many chiral centers does it have?

(b) How many optical isomers?

(c) Draw perspective formulas for all the optical isomers of isoleucine.

Answer

(a)   Two; at the α and β carbons (C-2 and C-3).

(b)   Four; the presence of two chiral centers produces four possible diastereoisomers: (S,S), (S,R), (R,R), and (R,S). Each of these, unlike enantiomers, is unique in its physical properties.

(c)

$$
\begin{array}{cccc}
\text{COO}^- & \text{COO}^- & \text{COO}^- & \text{COO}^- \\
\text{H}_3\overset{+}{\text{N}}\!-\!\overset{|}{\text{C}}\!-\!\text{H} & \text{H}_3\overset{+}{\text{N}}\!-\!\overset{|}{\text{C}}\!-\!\text{H} & \text{H}\!-\!\overset{|}{\text{C}}\!-\!\overset{+}{\text{N}}\text{H}_3 & \text{H}\!-\!\overset{|}{\text{C}}\!-\!\overset{+}{\text{N}}\text{H}_3 \\
\text{H}\!-\!\overset{|}{\text{C}}\!-\!\text{CH}_3 & \text{CH}_3\!-\!\overset{|}{\text{C}}\!-\!\text{H} & \text{H}\!-\!\overset{|}{\text{C}}\!-\!\text{CH}_3 & \text{CH}_3\!-\!\overset{|}{\text{C}}\!-\!\text{H} \\
\text{CH}_2 & \text{CH}_2 & \text{CH}_2 & \text{CH}_2 \\
\text{CH}_3 & \text{CH}_3 & \text{CH}_3 & \text{CH}_3
\end{array}
$$

9.   *Comparison of the pK_a Values of an Amino Acid and Its Peptides*   The titration curve of the amino acid alanine shows the ionization of two functional groups with pK_a values of 2.34 and 9.69, corresponding to the ionization of the carboxyl and the protonated amino groups, respectively. The titration of di-, tri-, and larger oligopeptides of alanine also shows the ionization of only two functional groups, although the experimental pK_a values are different. The trend in pK_a values is summarized in the table.

Amino acid or Peptide	pK_1	pK_2
Ala	2.34	9.69
Ala-Ala	3.12	8.30
Ala-Ala-Ala	3.39	8.03
Ala-(Ala)$_n$-Ala, n $\geq$ 4	3.42	7.94

(a)   Draw the structure of Ala-Ala-Ala. Identify the functional groups associated with pK_1 and pK_2.

(b)   The value of pK_1 *increases* in going from Ala to an Ala oligopeptide. Provide an explanation for this trend.

(c)   The value of pK_2 *decreases* in going from Ala to an Ala oligopeptide. Provide an explanation for this trend.

Answer

(a)

$$
\text{H}_3\overset{+}{\text{N}}\!-\!\underset{\underset{\text{CH}_3}{|}}{\text{CH}}\!-\!\overset{\overset{\text{O}}{\|}}{\text{C}}\!-\!\underset{\underset{\text{H}}{|}}{\text{N}}\!-\!\underset{\underset{\text{CH}_3}{|}}{\text{CH}}\!-\!\overset{\overset{\text{O}}{\|}}{\text{C}}\!-\!\underset{\underset{\text{H}}{|}}{\text{N}}\!-\!\underset{\underset{\text{NH}_3}{|}}{\text{CH}}\!-\!\overset{\overset{\text{O}}{\|}}{\text{C}}\!-\!\text{OH}
$$

$pK_2 = 9.69$ $pK_1 = 2.84$

Note that only the amino- and carboxyl-terminal groups ionize.

(b)   As the length of poly-Ala increases, the two terminal groups are moved farther apart, separated by an intervening sequence of "insulating" nonpolar structure. The proton on the terminal —COOH is lost at a higher pH, with greater difficulty (making the —COOH a weaker acid), since the positive charge of the —NH$_3^+$ is no longer nearby to "push it off" the proton or favor its loss.

(c)   The α-amino group becomes a weaker base, or stronger acid, losing its proton more easily from the zwitterionic form of poly-Ala than from Ala, since the negatively charged —COOH group is farther away. In Ala, the negative charge on the α-carboxyl group helps keep the proton bound to the α-amino group at higher pH values, an effect that is diminished with distance in poly-Ala.

10. **Peptide Synthesis**   In the synthesis of polypeptides on solid supports, the α-amino group of each new amino acid is "protected" by a *t*-butyloxycarbonyl group (see Box 5-2). What would happen if this protecting group were not present?

Answer   In the absence of a protecting or blocking group on the α-amino group, in the presence of the condensing agent, DCC, the amino acid being added to the growing chain would react not only with the growing chain but repeatedly with itself to produce a homopeptide, which might later attach itself to the growing chain in a nonstoichiometric fashion. Multiple unblocked amino acids might be added to the end of the growing chain. The efficiency of the addition of the amino acid to the growing polypeptide would be greatly reduced.

CHAPTER 6   Introduction to Proteins

1. *How Many β-Galactosidase Molecules Are Present in an* **E. coli** *Cell?*   *E. coli* is a rod-shaped bacterium 2 μm long and 1 μm in diameter. When grown on lactose (a sugar found in milk), the bacterium synthesizes the enzyme β-galactosidase (M_r 450,000), which catalyzes the breakdown of lactose. The average density of the bacterial cell is 1.2 g/mL, and 14% of its total mass is soluble protein, of which 1.0% is β-galactosidase. Calculate the number of β-galactosidase molecules in an *E. coli* cell grown on lactose.

> **Answer** Given that the cell density = 1.2 g/mL and that 1% of 14% of the cell mass is β-galactosidase, the enzyme concentration is
>
> $$\frac{(1.2 \text{ g/mL})(0.0014)}{450,000 \text{ g/mol}} = 3.7 \times 10^{-9} \text{ mol/mL} = 3.7 \times 10^{-6} \text{ mmol/mL}$$
>
> The volume of the cylindrical cell $= \pi r^2 h = (3.14)(0.5 \text{ μm})^2(2 \text{ μm}) = 1.57 \text{ μm}^3$
> $$= 1.57 \times 10^{-12} \text{ cm}^3 \approx 1.6 \times 10^{-12} \text{ mL}$$
>
> Multiplying concentration (mmol/L) by volume (mL) gives the amount of enzyme:
> $(1.6 \times 10^{-12} \text{ mL})(3.7 \times 10^{-6} \text{ mmol/mL}) = 5.9 \times 10^{-18} \text{ mmol}$
> $$= 5.9 \times 10^{-21} \text{ mol}$$
>
> The number of molecules of β-galactosidase is obtained by multiplying the number of moles by Avogadro's number:
> $(5.9 \times 10^{-21} \text{ mol})(6.02 \times 10^{23} \text{ molecules/mol}) = 3,500 \text{ molecules}$

2. *The Number of Tryptophan Residues in Bovine Serum Albumin*   A quantitative amino acid analysis reveals that bovine serum albumin contains 0.58% by weight of tryptophan, which has a molecular weight of 204.

    (a) Calculate the minimum molecular weight of bovine serum albumin (i.e., assuming there is only one tryptophan residue per protein molecule).

    (b) Gel filtration of bovine serum albumin gives a molecular weight estimate of about 70,000. How many tryptophan residues are present in a molecule of serum albumin?

Answer

(a) M_r of Trp in the polypeptide chain must be corrected for removal of water during peptide bond formation. For a Trp residue, $M_r = 204 - 18 = 186$.

The molecular weight of bovine serum albumin (BSA) can be calculated from % weight of a single residue by the following proportionality, where n is the number of Trp residues in the protein:

$$\frac{wt\ Trp}{wt\ BSA} = \frac{n \times M_r(Trp)}{M_r(BSA)}$$

$$\frac{0.58\ g}{100\ g} = \frac{n(186)}{M_r(BSA)}$$

Thus the minimum molecular weight ($n = 1$) of bovine serum albumin is

$$\frac{(100g)(186)(1)}{0.58g} = 32,100$$

(b)   The number of Trp residues, n, is $70,000/32,100 \approx 2.0$ per molecule of serum albumin.

3.   ***The Molecular Weight of Ribonuclease***   Lysine makes up 10.5% of the weight of ribonuclease. Calculate the minimum molecular weight of ribonuclease. The ribonuclease molecule contains ten lysine residues. Calculate the molecular weight of ribonuclease.

Answer   The M_r of a Lys redsidue $= 146 - 18 = 128$.
As in Problem 2, we can set up the proportionality:
$$\frac{10.5\ g}{100\ g} = \frac{n(128)}{M_r}$$

Thus, $M_r = n (1220)$, and 1,200 is the minimum molecular weight.
Since ribonuclease contains 10 Lys per molecule, $n = 10$ and $M_r = 12,200$.

4.   ***The Size of Proteins***   What is the approximate molecular weight of a protein containing 682 amino acids in a single polypeptide chain?

Answer   Assuming that the average M_r residue is 110 (corrected for loss of water), a protein containing 682 residues will have an M_r of approximately $682 \times 110 = 75,020$.

5.   ***Net Electric Charge of Peptides***   A peptide isolated from the brain has the sequence

Glu-His-Trp-Ser-Tyr-Gly-Leu-Arg-Pro-Gly

Determine the net charge on the molecule at pH 3. What is the net charge at pH 5.5? At pH 8? At pH 11? Estimate the pI for this peptide. (Use pK_a values for side chains and terminal amino and carboxyl groups as given in Table 5-1.)

Answer   When pH > pK_a, ionizing groups lose their protons. The pK_a values of importance here are those of the amino-terminal (2.3) and carboxyl-terminal (9.6) groups and those of the R groups of Glu (4.3), His (6.0), Tyr (10.1), and Arg (12.5).

pH	^+H_3N-Glu-	His-	Trp-	Ser-Tyr-	Gly-Leu-	Arg-	Pro-Gly-COO^-	Net charge
3.0	+1	0	+1	0		+1	-1	+2
5.5	+1	-1	+1	0		+1	-1	+1
8.0	+1	-1	0	0		+1	-1	0
11	0	-1	0	-1		+1	-1	-2

Two different methods can be used to estimate pI:

Average the pK_a values for the two side chains that ionize near pH 8, the amino-terminal α-amino group of Glu and the His imidazole group:

$$pI = \frac{(9.6 + 6.0)}{2} = 7.8$$

(A third group, Tyr-OH, ionizes near pH 10, and may contribute.)

Alternatively, plot the calculated net charges as a function of pH and determine graphically the pH at which the net charge is zero on the vertical axis. More data points are needed to use this method accurately.

6. **The Isoelectric Point of Pepsin**   Pepsin of gastric juice (pH ≈ 1.5) has a pI of about 1, much lower than that of other proteins (see Table 6-5). What functional groups must be present in relatively large numbers to give pepsin such a low pI? What amino acids can contribute such groups?

   Answer A low pI requires large numbers of negatively charged (low pK_a) carboxylate groups. These are contributed by Asp and Glu residues.

7. **The Isoelectric Point of Histones**   Histones are proteins of eukaryotic cell nuclei. They are tightly bound to deoxyribonucleic acid (DNA), which has many phosphate groups. The pI of histones is very high, about 10.8. What amino acids must be present in relatively large numbers in histones? In what way do these residues contribute to the strong binding of histones to DNA?

   Answer Large numbers of positively charged (high pK_a) groups give a high pI. These positively charged R groups, contributed by Lys, Arg, and His residues, interact strongly with the negatively charged phosphate groups of DNA through electrostatic (salt-bridge) bonds.

8. **Solubility of Polypeptides**   One method for separating polypeptides makes use of their differential solubilities. The solubility of large polypeptides in water depends upon the relative polarity of their R groups, particularly on the number of ionized groups: the more ionized groups there are, the more soluble the polypeptide. Which of each pair of polypeptides below is more soluble at the indicated pH?

   (a) $(Gly)_{20}$ or $(Glu)_{20}$ at pH 7.0

   (b) $(Lys-Ala)_3$ or $(Phe-Met)_3$ at pH 7.0

   (c) $(Ala-Ser-Gly)_5$ or $(Asn-Ser-His)_5$ at pH 6.0

   (d) $(Ala-Asp-Gly)_5$ or $(Asn-Ser-His)_5$ at pH 3.0

Answer

(a)   (Glu)$_{20}$; this polymer is highly negatively charged (ionic, polar) at pH 7. (Gly)$_{20}$ is uncharged except for the amino- and carboxyl-terminal groups.

(b)   (Lys-Ala)$_3$; this is highly positively charged (ionic, polar) at pH 7. (Phe-Met)$_3$ is much less polar and hence less soluble.

(c)   (Asn-Ser-His); both polymers have polar Ser side chains, but (Asn-Ser-His)$_5$ also has the neutral, polar Asn side chains and the partially protonated (ionized) His side chains, which makes it more soluble.

(d)   (Asn-Ser-His); at pH 3, the carboxylate side chains of Asp are partially protonated and neutral, whereas the imidazole groups of His are fully protonated and ionic.

9.   *Purification of an Enzyme*   A biochemist discovers and purifies a new enzyme, generating the purification table below:

Procedure	Total protein (mg)	Activity (units)
1. Crude extract	20,000	4,000,000
2. Precipitation (salt)	5,000	3,000,000
3. Precipitation (pH)	4,000	1,000,000
4. Ion exchange chromatography	200	800,000
5. Affinity chromatography	50	750,000
6. Size-exclusion chromatography	45	675,000

(a)   From the information given in the table, calculate the specific activity of the enzyme solution after each purification procedure.

Answer   From the percentage recovery of activity (units), we can calculate percentage yield and the specific activity (units/mg).

Procedure	Protein (mg)	Activity (units)	% Yield	Specific activity (units/mg)	Purification factor
1.	20,000	4,000,000	(100)	200	(1.0)
2.	5,000	3,000,000	75	600	x 3.0
3.	4,000	1,000,000	25	250	x 1.25
4.	200	800,000	20	4,000	x 20
5.	50	750,000	19	15,000	x 75
6.	45	675,000	17	15,000	x 75

(b)   Which of the purification procedures used for this enzyme is most effective (i.e., gives the greatest increase in purity)?

(c)   Which of the purification procedures is least effective?

(d)   Is there any indication in this table that the enzyme is now pure? What else could be done to estimate the purity of the enzyme preparation?

Answer

(b)   Step 4, ion-exchange chromatography, gives the greatest increase in specific activity, which is an index of purity and degree of increase in purification.

(c)   Step 3, pH precipitation, in which two-thirds of the total activity from the previous step were lost.

(d)   Yes; the specific activity was constant after step 5, affinity chromatography. SDS-polyacrylamide gel electrophoresis is an excellent, standard way of checking homogeneity and purity.

10.   *Fragmentation of a Polypeptide Chain by Proteolytic Enzymes*   Trypsin and chymotrypsin are specific enzymes that catalyze the hydrolysis of polypeptides at specific locations (Table 6-7). The sequence of the B chain of insulin is shown below. Note that the cystine cross-linkage between the A and B chains has been cleaved through the action of performic acid (see Fig. 6-12).

Phe-Val-Asn-Gln-His-Leu-CysSO$_3^-$-Gly-Ser-His-Leu-Val-Glu-Ala-Leu-
Tyr-Leu-Val-CysSO$_3^-$-Gly-Glu-Arg-Gly-Phe-Phe-Tyr-Thr-Pro-Lys-Ala

Indicate the points in the B chain that are cleaved by **(a)** trypsin and **(b)** chymotrypsin. Note that these proteases will not remove single amino acids from either end of a polypeptide chain.

Answer
(a) and (b) Trypsin (T) cleaves on the carbonyl-side of Lys and Arg residues; chymotrypsin (C) cleaves on the carbonyl-side of Phe, Tyr, and Trp residues. Thus, the cleavage pattern is (the amino-terminal end is at the left):

Phe-Val-Asn-Gln-His-Leu-CysSO$_3^-$-Gly-Ser-His-Leu-Val-Glu-Ala-Leu-

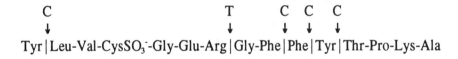

Tyr|Leu-Val-CysSO$_3^-$-Gly-Glu-Arg|Gly-Phe|Phe|Tyr|Thr-Pro-Lys-Ala

11.   *Sequence Determination of the Brain Peptide Leucine Enkephalin*   A group of peptides that influence nerve transmission in certain parts of the brain has been isolated from normal brain tissue. These peptides are known as opioids, because they bind to specific receptors that bind opiate drugs, such as morphine and naloxone. Opioids thus mimic some of the properties of opiates. Some researchers consider these peptides to be the brain's own pain killers. Using the information below, determine the amino acid sequence of the opioid leucine enkephalin. Explain how your structure is consistent with each piece of information.

(a)   Complete hydrolysis by 1 M HCl at 110 °C followed by amino acid analysis indicated by the presence of Gly, Leu, Phe, and Tyr, in a 2:1:1:1 molar ratio.

(b)   Treatment of the peptide with 1-fluoro-2,4-dinitrobenzene followed by complete hydrolysis and chromatography indicated the presence of the 2,4-dinitrophenyl derivative of tyrosine. No free tyrosine could be found.

(c)   Complete digestion of the peptide with pepsin followed by chromatography yielded a dipeptide containing Phe and Leu, plus a tripeptide containing Tyr and Gly in a 1:2 ratio.

Answer

(a) The empirical composition is (2Gly, Leu, Phe, Tyr)$_n$.

(b) Tyr is the amino-terminal residue, so the sequence is Tyr-(2Gly, Leu, Phe).

(c) Referring to Table 6-7, we see that pepsin cleaves on the amino-side of aromatic residues (Phe, Tyr, Trp). Since there are only two such residues in the peptide, one of which (Tyr) is amino-terminal, we can deduce that the dipeptide containing Phe and Leu was cleaved from the tripeptide at Phe. Thus the sequence must be:

$$\text{Tyr-(2Gly)-Phe-Leu} \;=\; \text{Tyr-Gly-Gly-Phe-Leu}$$

It is possible that $n > 1$ (i.e., that this sequence repeats). A good experimentalist would test this by a determination of M_r).

---

12. *Structure of a Peptide Antibiotic from* **Bacillus brevis** Extracts from the bacterium *Bacillus brevis* contain a peptide with antibiotic properties. Such peptide antibiotics form complexes with metal ions and apparently disrupt ion transport across the cell membrane, killing certain bacterial species. The structure of the peptide has been determined from the following observations.

(a) Complete acid hydrolysis of the peptide followed by amino acid analysis yielded equimolar amounts of Leu, Orn, Phe, Pro, and Val. Orn is ornithine, an amino acid not present in proteins but present in some peptides. It has the structure:

$$\overset{\displaystyle H}{H_3{}^+N-CH_2-CH_2-CH_2-\underset{\displaystyle {}^+NH_3}{\overset{\displaystyle |}{\underset{|}{C}}}-COO^-}$$

(b) The molecular weight of the peptide was estimated as about 1,200.

(c) When treated with the enzyme carboxypeptidase, the peptide failed to undergo hydrolysis.

(d) Treatment of the intact peptide with 1-fluoro-2,4-dinitrobenzene, followed by complete hydrolysis and chromatography, yielded only free amino acids and the following derivative:

$$O_2N-\!\!\!\bigcirc\!\!\!\overset{NO_2}{-}\!\!\!-NH-CH_2-CH_2-CH_2-\underset{{}^+NH_3}{\overset{H}{C}}-COO^-$$

(Hint: Note that the 2,4 dinitrophenyl derivative involves the amino group of a side chain rather than the α-amino group.)

(e) Partial hydrolysis of the peptide followed by chromatographic separation and sequence analysis yielded the di- and tripeptides below (the amino-terminal amino acid is always at the left):

Leu-Phe   Phe-Pro   Orn-Leu   Val-Orn
Val-Orn-Leu   Phe-Pro-Val   Pro-Val-Orn

Given the above information, deduce the amino acid sequence of the peptide antibiotic. Show your reasoning. When you have arrived at a structure, go back and demonstrate that it is consistent with *each* experimental observation.

Answer The information obtained from each experiment is as follows:

(a)   The simplest empirical formula for the peptide is (Leu, Orn, Phe, Pro, Val).

(b)   Assuming an average residue M_r of 110, the minimum molecular weight for the peptide is 550. Since $1200/550 \approx 2$, the empirical formula is (Leu, Orn, Phe, Pro, Val)$_2$.

(c)   Failure of carboxypeptidase to cleave the peptide could result from either the absence of a carboxyl-terminal residue or Pro at the carboxyl-terminus (which carboxypeptidase will not act upon).

(d)   Failure of FDNB to derivatize an α-amino group indicates either that there is no free amino-terminal group or that Pro (an imino acid) is at the amino-terminal position. (The derivative formed is 2,4 dinitrophenyl-ϵ-Lys.)

(e)   The presence of Pro at an internal position in the peptide Phe-Pro-Val indicates that it is *not* at the amino or carboxyl terminus. The information from these experiments allows us to conclude that the peptide is most likely cyclic.
Based on overlapping sequences:

$$
\begin{array}{l}
\text{Leu-Phe} \\
\quad\text{Phe-Pro} \\
\quad\text{Phe-Pro-Val} \\
\qquad\qquad\text{Pro-Val-Orn} \\
\qquad\qquad\text{Val-Orn} \\
\qquad\qquad\qquad\text{Orn-(Leu)}
\end{array}
$$

the most likely structure is a cyclic dimer of Leu-Phe-Pro-Val-Orn:

Leu-Phe-Pro-Val-Orn
↑   ↓
Orn-Val-Pro-Phe-Leu

where the arrows indicate the –CO–NH– or C→N direction of the peptide bonds.

CHAPTER 7

The Three-Dimensional Structure of Proteins

1. ***Properties of the Peptide Bond*** In x-ray studies of crystalline peptides Linus Pauling and Robert Corey found that the C—N bond in the peptide link is intermediate in length (0.132 nm) between a typical C—N single bond (0.149 nm) and a C=N double bond (0.127 nm). They also found that the peptide bond is planar (all four atoms attached to the C—N group are located in the same plane) and that the two α-carbon atoms attached to the C—N are always trans to each other (on opposite sides of the peptide bond):

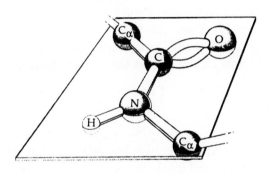

   (a) What does the length of the C—N bond in the peptide linkage indicate about its strength and its bond order, i.e., whether it is single, double, or triple?

   (b) In light of your answer to part (a), provide an explanation for the observation that such a C—N bond is intermediate in length between a double and single bond.

   (c) What do the observations of Pauling and Corey tell us about the ease of rotation about the C—N peptide bond?

   Answer
   (a) Shorter bonds are stronger than longer ones. The higher the bond order (multiple vs single), the shorter and stronger are the bonds. Thus bond length is an indication of bond order. For example, the C=N bond has a higher order ($n = 2.0$) and is shorter (0.127 nm) than a typical C—N bond (n = 1.0, length = 0.149 nm). The length of the C—N linkage in the peptide bond (0.132 nm) indicates that it is intermediate in strength and bond order between a single and double bond.

   (b) The peptide bond can be represented by two resonance structures (see Figure 7-4a), one single-bonded and the other double-bonded. The actual bond length is approximately the average of these two lengths.

(c)   Rotation about a double bond is generally impossible at physiological temperatures, and the steric relationship of the groups attached to the two atoms involved in the double bond is spatially "fixed." Since the peptide bond has considerable double-bond character, there is essentially no rotation and the C=O and N—H groups are fixed in the trans configuration.

2.   *Early Observations on the Structure of Wool* William Astbury discovered that the x-ray pattern of wool shows a repeating structural unit spaced about 0.54 nm along the direction of the wool fiber. When he steamed and stretched the wool, the x-ray pattern showed a new repeating structural unit at a spacing of 0.70 nm. Steaming and stretching the wool and then letting it shrink gave an x-ray pattern consistent with the original spacing of about 0.54 nm. Although these observations provided important clues to the molecular structure of wool, Astbury was unable to interpret them at the time. Given our current understanding of the structure of wool, interpret Astbury's observations.

Answer The principal structural units in the wool fiber polypeptides are successive turns of the α helix, which are spaced at 0.54 nm intervals. The intrinsic stability of the helix (and thus the fiber) results from *intra* chain hydrogen bonds (see Figure 7-13). Steaming and stretching the fiber yields an extended polypeptide chain with the β conformation, in which the distance between adjacent R groups is about 0.70 nm. Upon resteaming, the fiber reassumes an α-helical structure, the polypeptide chains assume the less extended α conformation, and the fiber shrinks.

3.   *Rate of Synthesis of Hair α-Keratin* In human dimensions, the growth of hair is a relatively slow process, occurring at a rate of 15 to 20 cm/yr. All this growth is concentrated at the base of the hair fiber, where α-keratin filaments are synthesized inside living epidermal cells and assembled into ropelike structures (see Fig. 7-13). The fundamental structural element of α-keratin is the α helix, which has 3.6 amino acid residues per turn and a rise of 0.56 nm per turn (see Fig. 7-6). Assuming that the biosynthesis of α-helical keratin chains is the rate-limiting factor in the growth of hair, calculate the rate at which peptide bonds of α-keratin chains must be synthesized (peptide bonds per second) to account for the observed yearly growth of hair.

Answer Since there are 3.6 amino acids (AAs) per turn and the rise is 0.56 nm/turn, the length per AA of the α helix is

$$\frac{0.56 \text{ nm—turn}}{3.6 \text{ AA/turn}} = 0.16 \text{ nm/AA}$$

$$= 1.6 \times 10^{-10} \text{ m/AA}$$

A growth rate of 20 cm/yr is equivalent to

$$\frac{20 \text{ cm/year}}{(365 \text{ days/yr})(24 \text{ h/day})(60 \text{ min/h})(60 \text{ s/min})} = 6.4 \times 10^{-7} \text{ cm/s} = 6.4 \times 10^{9} \text{ m/s}$$

Thus the rate at which AAs are added is

$$\frac{6.4 \times 10^{-9} \text{ m/s}}{1.6 \times 10^{-10} \text{ m/AA}} \approx 40 \text{ AA/s} \approx 40 \text{ peptide bonds per second}$$

4. *The Effect of pH on the Conformations of Polyglutamate and Polylysine*   The unfolding of the α helix of a polypeptide to a randomly coiled conformation is accompanied by a large decrease in a property called its specific rotation, a measure of a solution's capacity to rotate plane-polarized light. Polyglutamate, a polypeptide made up of only L-Glu residues, has the α-helical conformation at pH 3. However, when the pH is raised to 7, there is a large decrease in the specific rotation of the solution. Similarly, polylysine (L-Lys residues) is an α helix at pH 10, but when the pH is lowered to 7, the specific rotation also decreases, as shown by the following graph.

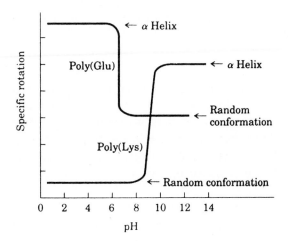

What is the explanation for the effect of the pH changes on the conformations of poly(Glu) and poly(Lys)? Why does the transition occur over such a narrow range of pH?

> *Answer* At pH values above 6, deprotonation of the carboxylate side chains of poly(Glu) leads to repulsion between adjacent negatively charged groups, which destabilizes the α helix and results in unfolding. Similarly, at pH 7 protonation of the amino-group side chains of poly(Lys) causes repulsion between positively charged groups that leads to unfolding.

5. *The Disulfide-Bond Content Determines the Mechanical Properties of Many Proteins*   A number of natural proteins are very rich in disulfide bonds, and their mechanical properties (tensile strength, viscosity, hardness, etc.) are correlated with the degree of disulfide bonding. For example, glutenin, a wheat protein rich in disulfide bonds, is responsible for the cohesive and elastic character of dough made from wheat flour. Similarly, the hard, tough nature of tortoise shell is due to the extensive disulfide bonding in its α-keratin. What is the molecular basis for the correlation between disulfide-bond content and mechanical properties of the protein?

> *Answer* Disulfide bonds are covalent bonds, which are much stronger than the noncovalent interactions (hydrogen bonds, hydrophobic interactions, van der Waals interactions) that stabilize the three-dimensional structure of most proteins. Disulfide bonds serve to cross-link protein chains, increasing stiffness, hardness, and mechanical strength.

6. *Why Does Wool Shrink?* When wool sweaters or socks are washed in hot water and/or dried in an electric dryer, they shrink. From what you know of α-keratin structure, how can you account for this? Silk, on the other hand, does not shrink under the same conditions. Explain.

> *Answer* Wool shrinks along the axis of the fiber under conditions of damp heat as the polypeptide chains in the extended-form coiled coil, protofibril, and microfibril within the α-keratin structure (see Figure 7-13) are converted to a β-conformation. See also the answer to Problem 2. (*Note*: the answer on p. AP-6 of the text is incorrect: this is *not a β-pleated sheet*). Under conditions of mechanical tension and moist heat, wool can be stretched back to a fully extended form. In contrast, the polypeptide chains of silk have a very stable β-pleated sheet structure, fully extended along the axis of the fiber (see Figure 7-9c), and have small, closely packed amino acid side chains. These characteristics make silk resistant to stretching or shrinking.

7. *Heat Stability of Proteins Containing Disulfide Bonds* Most globular proteins are denatured and lose their activity when briefly heated to 65 °C. Globular proteins that contain multiple disulfide bonds often must be heated longer at higher temperatures to denature them. One such protein is bovine pancreatic trypsin inhibitor (BPTI), which has 58 amino acid residues in a single chain and contains three disulfide bonds. On cooling a solution of denatured BPTI, the activity of the protein is restored. Can you suggest a molecular basis for this property?

> *Answer* As the temperature is raised, the increased thermal motion of the polypeptide chains and vibrational motions of hydrogen bonds ultimately lead to thermal denaturation (unfolding) of a protein. Cystine residues (disulfide bridges), depending on their location in the protein structure, can prevent or restrict the movement of folded protein domains, block access of solvent water to the interior of the protein, and prevent the complete unfolding of the protein. Refolding to the native structure from a random coil is seldom spontaneous, owing to the very large number of conformations possible. Disulfide bonds limit the number of conformations by allowing only a few minimally unfolded structures, and hence allow the protein to return to its native conformation more easily upon cooling.

8. *Bacteriorhodopsin in Purple Membrane Proteins* Under the proper environmental conditions, the salt-loving bacterium *Halobacterium halobium* synthesizes a membrane protein (M_r 26,000) known as bacteriorhodopsin, which is purple because it contains retinal. Molecules of this protein aggregate into "purple patches" in the cell membrane. Bacteriorhodopsin acts as a light-activated proton pump that provides energy for cell functions. X-ray analysis of this protein reveals that it consists of seven parallel α-helical segments, each of which traverses the bacterial cell membrane (thickness 4.5 nm). Calculate the minimum number of amino acids necessary for one segment of α helix to traverse the membrane completely. Estimate the fraction of the bacteriorhodopsin protein that occurs in α-helical form. Justify all your assumptions. (Use an average amino acid residue weight of 110.)

> *Answer* If there are seven helices in a protein of M_r 26,000 with an average residue M_r of 110, and assuming that all residues are involved in helices, the number of amino acids per helix is
>
> (26,000)/(7 helices/protein)(110 AA) = 34 AA/helix (maximum)

Using the parameters from Problem 3 (3.6 AA/turn, 0.56 nm/turn), we can calculate that there are 6.4 AA/nm along the axis of the helix. Thus in 4.5 nm of membrane, a minimum of (4.5 nm)(6.4 AA/nm) = 29 AAs are required to span the membrane per segment of helix.

Thus the fraction of residues involved in membrane-spanning helices is

29/34 = 0.85 = 85%

9. ***Biosynthesis of Collagen***   Collagen, the most abundant protein in mammals, has an unusual amino acid composition. Unlike most other proteins, collagen is very rich in proline and hydroxyproline (see p. 172). Hydroxyproline is not one of the 20 standard amino acids, and its incorporation in collagen could occur by one of two routes: (1) proline is hydroxylated by enzymes *before* incorporation into collagen or (2) a Pro residue is hydroxylated *after* incorporation into collagen. To differentiate between these two possibilities, the following experiments were performed. When [^{14}C]proline was administered to a rat and the collagen from the tail isolated, the newly synthesized tail collagen was found to be radioactive. If, however, [^{14}C]hydroxyproline was administered to a rat, no radioactivity was observed in the newly synthesized collagen. How do these experiments differentiate between the two possible mechanisms for introducing hydroxyproline into collagen?

> ***Answer***   Consider the two possible pathways for producing hydroxyproline (HO-Pro) residues:
>
> (1) Pro $\longrightarrow$ HO-Pro $\longrightarrow$ collagen(HO-Pro)
>
> (2) Pro $\longrightarrow$ collagen(Pro) $\longrightarrow$ collagen(HO-Pro)
>
> The observation that [^{14}C]-hydroxyproline is not incorporated into collagen argues against route (1), but is consistent with route (2). The modification of an amino acid residue after incorporation into a protein is called posttranslational modification.

10. ***Pathogenic Action of Bacteria That Cause Gas Gangrene***   The highly pathogenic anaerobic bacterium *Clostridium perfringens* is responsible for gas gangrene, a condition in which animal tissue structure is destroyed. This bacterium secretes an enzyme that efficiently catalyzes the hydrolysis of the peptide bond [indicated by an asterisk] in the sequence:

$$\overset{\text{\Large .}}{}\quad\quad\quad\overset{H_2O}{-X-Gly-Pro-Y- \longrightarrow -X-COO^- + {}^+H_3N-Gly-Pro-Y-}$$

where X and Y are any of the 20 standard amino acids. How does the secretion of this enzyme contribute to the invasiveness of this bacterium in human tissues? Why does this enzyme not affect the bacterium itself?

> ***Answer***   Collagen is distinctive in its amino acid composition having a very high proportion of Gly (35%) and Pro residues. The enzyme (a collagenase) secreted by the bacterium destroys the connective-tissue barrier (skin, hide, etc.) of the host, allowing the bacterium to invade the host tissues. Bacteria do not contain collagen and thus are unaffected by collagenase.

11. *Formation of Bends and Intrachain Cross-Linkages in Polypeptide Chains* In the following polypeptide, where might bends or turns occur? Where might intrachain disulfide cross-linkages be formed?

```
 1   2   3   4   5   6   7   8   9  10 11  12 13  14
Ile-Ala-His-Thr-Tyr-Gly-Pro-Phe-Glu-Ala-Ala-Met-Cys-Lys-

15  16   17  18  19  20  21  22 23  24 25  26 27  28
Trp-Glu-Ala-Gln-Pro-Asp-Gly-Met-Glu-Cys-Ala-Phe-His-Arg
```

> *Answer* Bends or turns are most likely to occur at residues 7 and 19, since Pro residues are often (not always) found at bends in globular, folded proteins. Other candidates are Ser, Thr, and Ile. Intrachain disulfide cross-linkages can form only between residues 13 and 24 (Cys residues).

12. *Location of Specific Amino Acids in Globular Proteins* X-ray analysis of the tertiary structure of myoglobin and other small, single-chain globular proteins has led to some generalizations about how the polypeptide chains of soluble proteins fold. With these generalizations in mind, indicate the probable location, whether in the interior or on the external surface, of the following amino acid residues in native globular proteins: Asp, Leu, Ser, Val, Gln, Lys. Explain your reasoning.

> *Answer* Examination of the tertiary structures of numerous proteins has provided a number of generalizations about how proteins fold. Amino acids with ionic (charged) or strongly polar neutral groups (such as Asp, Gln, and Lys) are located on the external surface to interact optimally with solvent water. In contrast, residues with nonpolar side chains (such as Val and Leu) are situated in the interior to escape the polar solvent environment. Ser is of intermediate polarity and can be found either in the interior or on the exterior surface (see Table 5-1).

13. *The Number of Polypeptide Chains in an Oligomeric Protein* A sample (660 mg) of an oligomeric protein of M_r 132,000 was treated with an excess of 1-fluoro-2,4-dinitrobenzene under slightly alkaline conditions until the chemical reaction was complete. The peptide bonds of the protein were then completely hydrolyzed by heating it with concentrated HCl. The hydrolysate was found to contain 5.5 mg of the following compound:

However, 2,4-dinitrophenyl derivatives of the α-amino groups of other amino acids could not be found.

(a) Explain why this information can be used to determine the number of polypeptide chains in an oligomeric protein.

(b) Calculate the number of polypeptide chains in this protein.

Answer

(a) Since only a single 2,4-dinitrophenyl (DNP) amino acid derivative is found, there is only one kind of amino acid at the amino terminus, i.e., all of the polypeptide chains have the same amino-terminal residue. Comparing the number of moles of this derivative to the number of moles of protein gives the number of polypeptide chains.

(b) The amount of protein = (0.66 g)/(132,000 g/mol) = 5 x 10^{-6} mol.
Since M_r for DNP-Val ($C_{11}H_{13}O_6N_3$) = 283,
the amount of DNP-Val = (0.0055 g)/(283 g/mol) = 19.4 x 10^{-6} mol.
The ratio of moles of DNP-Val to moles of protein gives the number of amino-terminal residues and thus the number of chains per holoenzyme:

$$\frac{1.9 \times 10^{-5} \text{ mol DNP-Val}}{5 \times 10^{-6} \text{ mol protein}} = 3.8 \approx 4 \text{ polypeptide chains}$$

An alternative approach to the problem is through the proportionality:

$$\frac{n(283)}{132,000} = \frac{5.5 \text{ mg}}{660 \text{ mg}}$$

$$n = \frac{(5.5 \text{ mg})(132,000)}{(660 \text{ mg})(283)} = 3.89 \approx 4$$

14. *Molecular Weight of Hemoglobin* The first indication that proteins have molecular weights greatly exceeding those of the (then known) organic compounds was obtained over 100 years ago. For example, it was known at that time that hemoglobin contains 0.34% by weight of iron.

(a) From this information determine the minimum molecular weight of hemoglobin.

(b) Subsequent experiments indicated that the true molecular weight of hemoglobin is 64,500. What information did this provide about the number of iron atoms in hemoglobin?

Answer These calculations are similar to those in Chapter 6, Problems 2 and 3.
(a)

$$\frac{0.34 \text{ g}}{100 \text{ g}} = \frac{n(55.85)}{M_r}$$

For $n = 1$, the minimum molecular weight is

$$\frac{100 \text{ g}(55.85)}{0.34 \text{ g}} = 16,426 \approx 16,400$$

(b) The number of iron atoms, n, is 64,500/16,400 = 3.9 ≈ 4.0 per hemoglobin molecule.

15. *Comparison of Fetal and Maternal Hemoglobin*   Studies of oxygen transport in pregnant mammals have shown that the O_2-saturation curves of fetal and maternal blood are markedly different when measured under the same conditions. Fetal erythrocytes contain a structural variant of hemoglobin, hemoglobin F, consisting of two α and two γ subunits ($\alpha_2\gamma_2$), whereas maternal erythrocytes contain the usual hemoglobin A ($\alpha_2\beta_2$).

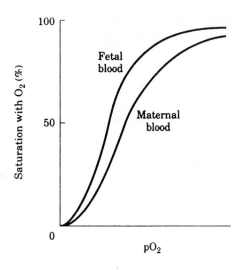

(a) Which hemoglobin has a higher affinity for oxygen under physiological conditions, hemoglobin A or hemoglobin F? Explain.

(b) What is the physiological significance of the different oxygen affinities? Explain.

Answer
(a) Hemoglobin F; its pO_2 (the pressure of oxygen required for half-saturation) is lower than that of HbA, thus HbF binds oxygen better than does HbA at low oxygen levels. An alternative way of looking at this is that at pO_2 = 4kPa, HbF is 58% saturated, whereas HbA is only 33% saturated. Both comparisons indicate that HbF binds oxygen more efficiently than HbA under physiological conditions.

(b) The higher oxygen affinity (lower pO_2 for half-saturation) of HbF means that oxygen passes (or is pulled) from adult (maternal) blood to fetal blood in the placenta. To optimize the transport of oxygen, the pO_2 should be near or below the half-saturation point for HbA. This is in fact the case.

CHAPTER **8**   **Enzymes**

1. ***Keeping the Sweet Taste of Corn***   The sweet taste of freshly picked corn is due to the high level of sugar in the kernels. Store-bought corn (several days after picking) is not as sweet, because about 50% of the free sugar of corn is converted into starch within one day of picking. To preserve the sweetness of fresh corn, the husked ears are immersed in boiling water for a few minutes ("blanched") and then cooled in cold water. Corn processed in this way and stored in a freezer maintains its sweetness. What is the biochemical basis for this procedure?

     Answer   After an ear of corn has been removed from the plant, the enzymatic conversion of sugar to starch continues. Inactivation of the enzymes responsible slows down this conversion to an imperceptible rate. One of the simplest techniques is heat denaturation. Freezing the corn lowers any remaining enzyme activity to an insignificant level.

2. ***Intracellular Concentration of Enzymes***   To approximate the actual concentration of enzymes in a bacterial cell, assume that the cell contains 1,000 different enzymes in solution in the cytosol, that each protein has a molecular weight of 100,000, and that all 1,000 enzymes are present in equal concentrations. Assume that the bacterial cell is a cylinder (diameter 1 μm, height 2.0 μm). If the cytosol (specific gravity 1.20) is 20% soluble protein by weight, and if the soluble protein consists entirely of different enzymes, calculate the *average* molar concentration of each enzyme in this hypothetical cell.

     Answer   There are three different ways to approach this problem.

     (i)   The concentration of total protein in the cell is

$$\frac{(1.2 \text{ g/mL})(0.2)}{100,000 \text{ g/mol}} = 0.24 \times 10^{-5} \text{ mol/mL} = 2.4 \times 10^{-3} \text{ M}$$

Thus, for one enzyme in 1,000, the enzyme concentration $= \dfrac{2.4 \times 10^{-3} \text{ M}}{(1000)}$

$$= 2.4 \times 10^{-6} \text{ mol/L}$$
$$\approx 2 \ \mu\text{M}$$

     (ii)   The average molar concentration $= \dfrac{\text{moles of each enzyme in cell}}{\text{volume of cell in liters}}$

Volume of bacterial cytosol $= \pi r^2 h = (3.14)(0.5)^2(2) \ \mu\text{m}^3 = 1.57 \ \mu\text{m}^3$
$$= 1.57 \times 10^{-15} \text{ L}$$

Amount (in moles) of each enzyme in cell is

$$\frac{(0.20)(1.2 \text{ g/cm}^3)(1.57 \ \mu\text{m}^3)(10^{-12}\text{cm}^3/\mu\text{m}^3)}{(100,000 \text{ g/mol})(100,000)} = 3.77 \times 10^{-21} \text{ mol}$$

Average molar concentration $= \dfrac{3.77 \times 10^{-21} \text{ mol}}{1.57 \times 10^{-15} \text{ L}}$

$$= 2.4 \times 10^{-6} \text{ mol/L} \approx 2 \ \mu\text{M}$$

(iii)   Volume of bacterial cytosol $= \pi r^2 h$
$$= (3.14)(0.5)^2(2) \ \mu\text{m}^3 = 1.57 \ \mu\text{m}^3 = 1.57 \times 10^{-12} \text{ mL}$$
Wt. of cytosol $=$ (specific gravity)(volume)
$$= (1.2 \text{ g/mL})(1.57 \times 10^{-12} \text{ mL}) = 1.88 \times 10^{12} \text{ g}$$
Average weight of each protein (1 in 1000, 20% wt/wt protein)
$$= (1.88 \times 10^{-12} \text{ g})(0.2)/(1000) = 3.77 \times 10^{-13} \text{ g}$$
Average molar concentration of each protein
$$= \text{(average weight)}/(M_r)\text{(volume)}$$
$$= (3.77 \times 10^{-13} \text{ g})(10^5 \text{ g/mol})(1.57 \times 10^{-12} \text{ mL})(1000 \text{ mL/L})$$
$$= 2.4 \times 10^{-6} \text{ mol/L} \approx 2 \ \mu\text{M}$$

3. *Rate Enhancement by Urease*   The enzyme urease enhances the rate of urea hydrolysis at pH 8.0 and 20 °C by a factor of 10^{14}. If a given quantity of urease can completely hydrolyze a given quantity of urea in 5 min at 20 °C and pH 8.0, how long will it take for this amount of urea to be hydrolyzed under the same conditions in the absence of urease? Assume that both reactions take place in sterile systems so that bacteria cannot attack the urea.

   Answer

$$\frac{(5 \text{ min})(10^{14})}{(60 \text{ min/h})(24 \text{ hr/day})(365 \text{ days/yr})}$$

$$= 9.5 \times 10^8 \text{ yr}$$
$$= 950 \text{ million years!}$$

4. *Requirements of Active Sites in Enzymes*   The active site of an enzyme usually consists of a pocket on the enzyme surface lined with the amino acid side chains necessary to bind the substrate and catalyze its chemical transformation. Carboxypeptidase, which sequentially removes the carboxyl-terminal amino acid residues from its peptide substrates, consists of a single chain of 307 amino acids. The two essential catalytic groups in the active site are furnished by Arg^{145} and Glu^{270}.

   (a) If the carboxypeptidase chain were a perfect α helix, how far apart (in nanometers) would Arg^{145} and Glu^{270} be? (Hint: See Fig. 7-6.)

   (b) Explain how it is that these two amino acids, so distantly separated in the sequence, can catalyze a reaction occurring in the space of a few tenths of a nanometer.

   (c) If only these two catalytic groups are involved in the mechanism of hydrolysis, why is it necessary for the enzyme to contain such a large number of amino acid residues?

Answer

(a) Arg^{145} is separated from Glu^{270} by (270 - 145) = 125 amino acid (AA) residues. From Fig. 7-6, we see that the α helix has 3.6 AA/turn and increases in length along the major axis by 0.56 nm/turn. Thus, the distance between the two residues is

$$\frac{(125 \text{ AA})(0.56 \text{ nm/turn})}{3.6 \text{ AA/turn}} = 18.8 \text{ nm}$$

(b) Three-dimensional folding of the enzyme brings the two amino acid residues into close proximity.

(c) In principle, one could envision a short peptide containing the two amino acids in close proximity. In actual practice, such a short peptide would fail to fold properly. Thus one of the reasons for such a long linear sequence is that the folding of the protein assures the correct orientation of catalytic groups in the active site. Furthermore, experiments indicate that the protein serves as a "scaffolding" to keep the catalytic groups in a precise orientation. Also, many other (noncatalytic) interactions occur between the enzyme and its substrate, and some of the binding energy derived from these interactions contributes to catalysis.

5. *Quantitative Assay for Lactate Dehydrogenase* The muscle enzyme lactate dehydrogenase catalyzes the reaction

$$CH_3-CO-COO^- + NADH + H^+ \longrightarrow CH_3-CH(OH)-COO^- + NAD^+$$
      Pyruvate   Lactate

NADH and NAD^+ are the reduced and oxidized forms, respectively, of the coenzyme NAD. Solutions of NADH, but *not* NAD^+, absorb light at 340 nm. This property is used to determine the concentration of NADH in solution by measuring spectrophotometrically the amount of light absorbed at 340 nm by the solution. Explain how these properties of NADH can be used to design a quantitative assay for lactate dehydrogenase.

Answer The reaction rate can be measured by following the decrease in NADH absorption at 340 nm as the reaction proceeds. Three pieces of information are required to develop a good quantitative assay for LDH:

(i) Determine K_m values (see Box 8-1);

(ii) Measure the initial rate at several known concentrations of enzyme with saturating concentrations of NADH and pyruvate; and

(iii) Plot the initial rates as a function of [E]. This plot should be linear, with a slope that provides a measure of LDH concentration.

6. *Estimation of* V_{max} *and* K_m *by Inspection* Although graphical methods are available for accurate determination of the values of V_{max} and K_m of an enzyme-catalyzed reaction (see Box 8-1), these values can be quickly estimated by inspecting values of V_o at increasing [S]. Estimate the approximate value of V_{max} and K_m for the enzyme-catalyzed reaction for which the following data were obtained:

[S] (M)	V_0 (μM/min)	[S] (M)	V_0 (μM/min)
2.5×10^{-6}	28	4×10^{-5}	112
4.0×10^{-6}	40	1×10^{-4}	128
1×10^{-5}	70	2×10^{-3}	139
2×10^{-5}	95	1×10^{-2}	140

Answer To estimate V_{max}, notice how little the velocity changes as the substrate concentration increases by fivefold from 2 to 10 mM. Thus, we can estimate that $V_{max} \approx 140$ μM/min.

K_m is defined as the substrate concentration that produces $\frac{1}{2}V_{max}$, or 70 μM/min. Inspection of the table indicates that this V_0 value occurs at [S] = 1×10^{-5} M $\approx K_m$.

7. *Relation between Reaction Velocity and Substrate Concentration: Michaelis-Menten Equation*
    (a) At what substrate concentration will an enzyme having a k_{cat} of 30 s^{-1} and a K_m of 0.005 M show one-quarter of its maximum rate?

    (b) Determine the fraction of V_{max} that would be found in each case when [S] = $\frac{1}{2}K_m$, $2K_m$, and $10K_m$.

    Answer
    (a) Since $V_0 = V_{max}[S]/(K_m + [S])$, and $V_0 = 0.25(30$ s$^{-1})$, = 7.5 s^{-1}, we can substitute into the Michaelis-Menten equation to give

    7.5 s^{-1} = (30 s^{-1} [S]/(5 mM + [S])

    [S] = 1.7 mM = 1.7×10^{-3} M

    (b) We can rearrange the Michaelis-Menten equation into the form
    $$V_0/V_{max} = [S]/(K_m + [S])$$

    Substituting [S] = $\frac{1}{2}K_m$ into this equation gives
    $V_0/V_{max} = 0.33$
    Substituting [S] = $2 K_m$ into the equation and gives
    $V_0/V_{max} = 0.67$
    Finally, substituting [S] = $10K_m$ into the equation gives
    $V_0/V_{max} = 0.91$

8. *Graphical Analysis of* V_{max} *and* K_m *Values*   The following experimental data were collected during a study of the catalytic activity of an intestinal peptidase capable of hydrolyzing the dipeptide glycylglycine:

$$\text{Glycylglycine} + H_2O \longrightarrow 2 \text{ glycine}$$

[S](mM)	Product formed (μmol/min)	[S](mM)	Product formed (μmol/min)
1.5	0.21	4.0	0.33
2.0	0.24	8.0	0.40
3.0	0.28	16.0	0.45

From these data determine by graphical analysis (see Box 8-1) the values of K_m and V_{max} for this enzyme preparation and substrate.

Answer   As described in Box 8-1, the standard method is to use the V_o versus [S] data to calculate $1/V_o$ and $1/[S]$.

V_o (mg/min)	$1/V_o$ (min/mg)	[S] (mM)	$1/[S]$ (mM^{-1})
0.21	4.76	1.5	0.67
0.24	4.17	2.0	0.50
0.28	3.57	3.0	0.33
0.33	3.03	4.0	0.25
0.40	2.50	8.0	0.13
0.45	2.22	16.0	0.06

Plotting these reciprocal values gives a Lineweaver-Burk plot. From the best straight line through the data, the intercept on the horizontal axis gives the value $-1/K_m$ and the intercept on the vertical axis gives $1/V_{max}$. From these values, we can calculate:

$$V_{max} = 0.51 \ \mu\text{mol/min}$$
$$K_m = 2.2 \text{ mM}$$

9. *The Turnover Number of Carbonic Anhydrase*   Carbonic anhydrase of erythrocytes (M_r 30,000) is among the most active of known enzymes. It catalyzes the reversible hydration of CO_2:

$$H_2O + CO_2 \rightleftharpoons H_2CO_3$$

which is important in the transport of CO_2 from the tissues to the lungs.

(a) If 10 μg of pure carbonic anhydrase catalyzes the hydration of 0.30 g of CO_2 in 1 min at 37 °C under optimal conditions, what is the turnover number (k_{cat}) of carbonic anhydrase (in units of min^{-1})?

Answer The turnover number of an enzyme is defined as the number of substrate molecules transformed per unit time by a single enzyme molecule (or a single catalytic site) when the enzyme is saturated with substrate:

$k_{cat} = V_{max}/E_t$

where E_T = total moles of active sites.

We can convert the values given in the problem into a turnover number (min^{-1}) by converting the weights of enzyme and substrate to molar amounts:

V_{max}(moles of CO_2/min) $= \dfrac{0.3 \text{ g/min}}{44 \text{ g/mol}} = 6.8 \times 10^3$ mol/min

Amount of enzyme (moles) $= \dfrac{(10\ \mu g)(1\ g/10^6\ \mu g)}{30{,}000 \text{ g/mol}} = 3.3 \times 10^{-10}$ mol

The turnover number is obtained by dividing moles of CO_2/min by moles of enzyme:

$k_{cat} = \dfrac{6.8 \times 10^3 \text{ mol/min}}{3.3 \times 10^{-10} \text{ mol}} = 2.0 \times 10^7 \text{ min}^{-1}$

(b) From the answer in (a), calculate the activation energy of the enzyme-catalyzed reaction (in kJ/mol).

Answer Eqn. 8-6 (p. 204) can be rearranged and written as

$\Delta G\ddagger = -RT \ln k$

where k has units of s^{-1}.
Dividing by (60 sec/min) , the turnover number from (a) can be converted to a value of $3.3 \times 10^{-5}\ s^{-1}$. Thus
$\Delta G\ddagger = -(8.32\ \text{J/mol·K})(310\ \text{K})(2.3) \log (3.3 \times 10^{-5}\ s^{-1}) = 32.7$ kJ/mol

(*Note*: the value given on p. AP-6 of the text (43 kJ/mol) is obtained from $k_{cat} = 2 \times 10^7\ \text{min}^{-1}$, i.e., using incorrect units; note also that ln K in the equation is taken as being unitless.

(c) If carbonic anhydrase provides a rate enhancement of 10^7, what is the activation energy for the uncatalyzed reaction?

Answer Each rate enhancement factor of 10 is equivalent to 5.7 kJ of activation energy (see p. 207 of text), so 10^7 is equivalent to 7(5.7 kJ) = 40 kJ.
(Note: the value in the text, 83.2 kJ, is incorrect; see note at end of (b).)

10. *Irreversible Inhibition of an Enzyme* Many enzymes are inhibited irreversibly by heavy-metal ions such as Hg^{2+}, Cu^{2+}, or Ag^+, which can react with essential sulfhydryl groups to form mercaptides:

$$\text{Enz}-\text{SH} + \text{Ag}^+ \longrightarrow \text{Enz}-\text{S}-\text{Ag} + \text{H}^+$$

The affinity of Ag^+ for sulfhydryl groups is so great that Ag^+ can be used to titrate —SH groups quantitatively. To 10 mL of a solution containing 1.0 mg/mL of a pure enzyme was added just enough $AgNO_3$ to completely inactivate the enzyme. A total of 0.342 μmol of $AgNO_3$ was required. Calculate the *minimum* molecular weight of the enzyme. Why does the value obtained in this way give only the minimum molecular weight?

Answer An equivalency exists:

$$0.342 \times 10^{-3} \text{ mmol} = \frac{(10\text{mg/mL})(10\text{mL})}{(\text{minimum } M_r) \text{ mg/mmol}}$$

Thus minimum $M_r = \dfrac{(10 \text{ mg/mL})(10\text{mL})}{0.342 \times 10^{-3} \text{ mmol}} = 2.9 \times 10^4 = 29,000$

It is assumed that the enzyme contains only one titratable —SH group per molecule.

11. *Protection of an Enzyme against Denaturation by Heat* When enzyme solutions are heated, there is a progressive loss of catalytic activity with time. This loss is the result of the unfolding of the native enzyme molecule to a randomly coiled conformation, because of its increased thermal energy. A solution of the enzyme hexokinase incubated at 45 °C lost 50% of its activity in 12 min, but when hexokinase was incubated at 45 °C in the presence of a very large concentration of one of its substrates, it lost only 3% of its activity. Explain why thermal denaturation of hexokinase was retarded in the presence of one of its substrates.

> *Answer* One possibility is that the E-S complex is more stable than free enzyme. This implies that the ground state for the E-S complex is lower than that for free enzyme, which *increases the height of the energy barrier* to be crossed in reaching the denatured or unfolded state.
>
> An alternative view is that enzymes undergo unfolding in two stages:
>
> $$N \rightharpoonup U \Longleftrightarrow I$$
>
> involving reversible conversion of active native enzyme, N, to an unfolded, inactive state, U, followed by irreversible conversion to I. If substrate binds only to N, then saturation with S to form N-S complexes makes less free N available for conversion to U or I, as the N $\rightharpoonup$ U equilibrium is perturbed toward N. If N, but not N-S complexes, are converted to U or I, this will cause stabilization.

12. *Clinical Application of Differential Enzyme Inhibition* Human blood serum contains a class of enzymes known as acid phosphatases, which hydrolyze biological phosphate esters under slightly acidic conditions (pH 5.0):

$$R-O-P-O_3^{2-} + H_2O \longrightarrow R-OH + HO-P-O_3^{2-}$$

Acid phosphatases are produced by erythrocytes, the liver, kidney, spleen, and prostate gland. The enzyme from the prostate gland is clinically important because an increased activity in the blood is frequently an indication of cancer of the prostate gland. The phosphatase from the prostate gland is strongly inhibited by the tartrate ion, but acid phosphatases from other tissues are not. How can this information be used to develop a specific procedure for measuring the activity of the acid phosphatase of the prostate gland in human blood serum?

> *Answer* First measure the *total* acid phosphatase activity in human blood serum in units of μmol of phosphate ester hydrolyzed/mL of serum. Then, remeasure this activity in the presence of tartrate ion at a concentration sufficient to completely inhibit the enzyme from prostate gland. The difference between these two activities represents the acid phosphatase activity from the prostate gland.

13. ***Inhibition of Carbonic Anhydrase by Acetazolamide***   Carbonic anhydrase is strongly inhibited by the drug acetazolamide, which is used as a diuretic (increases the production of urine) and to treat glaucoma (reduces excessively high pressure within the eyeball). Carbonic anhydrase plays an important role in these and other secretory processes, because it participates in regulating the pH and bicarbonate content of a number of body fluids. The experimental curve of reaction velocity (given here as percentage of V_{max}) versus [S] for the carbonic anhydrase reaction is illustrated below (upper curve). When the experiment is repeated in the presence of acetazolamide, the lower curve is obtained. From an inspection of the curves and your knowledge of the kinetic properties of competitive and noncompetitive enzyme inhibitors, determine the nature of the inhibition by acetazolamide. Explain.

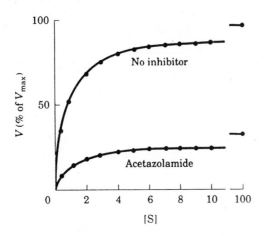

Answer   The graph gives us several pieces of information. First, the inhibitor prevents the enzyme from achieving the same V_{max} as in the absence of inhibitor. Second, the overall shape of the two curves is very similar: at any [S] the ratio of the two velocities ($\pm$ inhibitor) is the same. Third, the velocity does not change very much above [S] = 10, so at [S] = 100, the observed velocity is essentially V_{max} on each curve. Fourth, if we estimate the [S] at which ½ V_{max} is achieved, these values are nearly identical for both curves. Noncompetitive inhibitors alter V_{max} of enzymes but leave K_m unchanged. Thus, acetazolamide acts as a noncompetitive inhibitor of carbonic anhydrase.

14. *pH Optimum of Lysozyme*   The enzymatic activity of lysozyme is optimal at pH 5.2 (see graph below).   The active site of lysozyme contains two amino acid residues essential for catalysis: Glu^{35} and Asp^{52}.   The pK_a values of the carboxyl side chains of these two residues are 5.9 and 4.5, respectively.   What is the ionization state (protonated or deprotonated) of each residue at the pH optimum of lysozyme?   How can the ionization states of these two amino acid residues explain the pH-activity profile of lysozyme shown below?

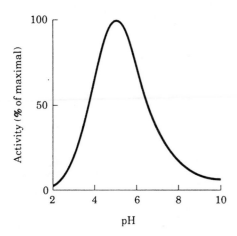

Answer   At a pH midway between the two pK_a values (pH 5.2), the side chain carboxyl group of Asp^{52} (with the lower pK_a = 4.5) will be mainly deprotonated (—COO$^-$), whereas Glu^{35} (with the higher pK_a = 5.9, the stronger base) will be protonated (—COOH).   At pH values below 5.2, Asp^{52} becomes protonated and the activity decreases.   Similarly, at pH values above 5.2, Glu^{35} becomes deprotonated and the activity also decreases.   The pH-activity profile suggests that maximum catalytic activity occurs at a pH midway between the pK_a values of the two acidic groups, when Glu^{35} is protonated and Asp^{52} is deprotonated.

CHAPTER 9 Lipids

1. *Melting Points of Fatty Acids* The melting points of a series of 18-carbon fatty acids are stearic acid, 69.6 °C; oleic acid, 13.4 °C; linoleic acid, -5 °C; and linolenic acid, -11 C. What structural aspect of these 18-carbon fatty acids can be correlated with the melting point? Provide a molecular explanation for the trend in melting points.

    Answer The number of cis double bonds. Each cis double bond causes a bend in the hydrocarbon chain; it is more difficult to pack these bent chains in the crystal lattice. The lower the extent of packing, the lower the melting temperature.

2. *Spoilage of Cooking Fats* Some fats used in cooking, such as olive oil, spoil rapidly upon exposure to air at room temperature, whereas others, such as solid shortening, remain unchanged. Why?

    Answer Unsaturated fats, such as vegetable oil, are susceptible to oxidation by molecular oxygen to form shorter chain aldehydes and acids, thus becoming "rancid." Saturated fats, lacking olefinic bonds, are not as reactive with oxygen.

3. *Preparation of Béarnaise Sauce* During the preparation of béarnaise sauce, egg yolks are incorporated into melted butter to stabilize the sauce and avoid separation. The stabilizing agent in the egg yolks is lecithin (phosphatidylcholine). Suggest why this works.

    Answer Lecithin serves as an emulsifying agent. The amphipathic nature of phosphatidylcholine allows it to facilitate the solubilization of butter.

4. *Hydrolysis of Lipids* Name the products of mild hydrolysis of the following lipids with dilute NaOH:
    (a) 1-stearoyl-2,3-dipalmitoylglycerol
    (b) 1-palmitoyl-2-oleoylphosphatidylcholine

    Answer Mild hydrolysis breaks the ester linkages between glycerol and fatty acids, forming:

    (a) the sodium salt of palmitic and stearic acids, plus glycerol.

    (b) the sodium salts of palmitic and oleic acids, plus glycerol-3-phosphorylcholine.

5. *Number of Detergent Molecules per Micelle* When a small amount of sodium dodecyl sulfate ($Na^+CH_3(CH_2)_{11}OSO_3^-$) is dissolved in water, the detergent ions go into solution as monomeric species. As more detergent is added, a point is reached (the critical micelle concentration) at which the monomers associate to form micelles. The critical micelle concentration of SDS is 8.2 mM. An examination of the micelles shows that they have an average particle weight (the sum of the molecular weights of the constituent monomers) of 18,000. Calculate the number of detergent molecules in the average micelle.

> *Answer* The molecular weight of sodium dodecyl sulfate is 288. Given an average micelle particle weight of 18,000, there are 18,000/288 = 63 SDS molecules per micelle.

6. *Hydrophobic and Hydrophilic Components of Membrane Lipids* A common structural feature of membrane lipid molecules is their amphipathic nature. For example, in phosphatidylcholine, the two fatty acid chains are hydrophobic and the phosphocholine head group is hydrophilic. For each of the following membrane lipids, name the components that serve as the hydrophobic and hydrophilic units:

   (a) phosphatidylethanolamine
   (b) sphingomyelin
   (c) galactosylcerebroside
   (d) ganglioside
   (e) cholesterol

   Answer

	Hydrophobic unit(s)	Hydrophilic unit(s)
(a)	2 Fatty acids	Phosphoethanolamine
(b)	1 Fatty acid and the hydrocarbon chain of spingosine	Phosphocholine
(c)	1 Fatty acid and the hydrocarbon chain of spingosine	D-Galactose
(d)	1 Fatty acid and the hydrocarbon chain of spingosine	Several sugar molecules
(e)	Hydrocarbon backbone	Alcohol group

7. *Properties of Lipids and Lipid Bilayers* Lipid bilayers formed between two aqueous phases have this important property: they form two-dimensional sheets, the edges of which close upon each other, and undergo self-sealing to form liposomes.

   (a) What properties of lipids are responsible for this property of bilayers? Explain.
   (b) What are the biological consequences of this property with regard to the structure of biological membranes?

   Answer
   (a) The amphipathic nature of lipids contributes to their ability to form liposomes. Formation of the lipid bilayer of liposomes minimizes the interaction between water and the hydrophobic side chains of the lipid. The driving force is entropy, which is increased by minimizing the number of water molecules that are more ordered at the lipid-water interface.

(b) The biological consequences of this property of lipids are formation of membranes. This allows for the formation and compartmentation of a cell.

8. *Chromatographic Separation of Lipids* A mixture of the following lipids is applied to a silica gel column, and the column is then washed with progressively more polar solvents. The mixture consists of: phosphatidylserine, cholesteryl palmitate (a sterol ester), phosphatidylethanolamine, phosphatidylcholine, sphingomyelin, palmitic acid, *n*-tetradecanol, triacylglycerol, and cholesterol. In what order do you expect the lipids to elute from the column?

Answer Because silica gel is polar, the order of elution occurs from the most hydrophobic to the most hydrophilic. The neutral lipids would be eluted first: cholesteryl palmitate and triacylglycerol. Because of their somewhat higher polarity, cholesterol and *n*-tetradecanol, although neutral, would elute after the neutral lipids. Sphingomyelin would elute next, before the phospholipids: phosphatidylserine, phosphatidylethanolamine, and phosphatidylcholine. Palmitic acid, the most polar lipid in the mixture, would elute last.

9. *Storage of Fat-Soluble Vitamins* In contrast to water-soluble vitamins, which must be a part of our daily diet, fat-soluble vitamins can be stored in the body in amounts sufficient for many months. Suggest an explanation for this difference based on solubilities.

Answer In contrast to water-soluble compounds, lipid-soluble compounds are not readily mobilized. Therefore, the body's lipids provide a reservoir for storage of lipid-soluble vitamins. Water-soluble vitamins are rapidly excreted by the kidneys.

10. *Alkali Lability of Triacylglycerols* A common procedure for cleaning the grease trap in a sink is to add a product that contains sodium hydroxide. Explain why this works.

Answer Triacylglycerols, a common component of the grease that plugs sinks, are hydrolyzed by NaOH to form the sodium salts of free fatty acids and glycerol, a process known as saponification. The fatty acids form micelles, which are more water soluble than triacylglycerols.

11. *Dependence of Melting Point on Fatty Acid Unsaturation* Draw all of the possible triacylglycerols that you could construct from glycerol, palmitic acid, and oleic acid. Rank them in order of increasing melting point.

Answer Six different triacylglycerols can be constructed. (*Note*: not eight, as on p. AP-7 of the text.) All saturated (palmitic) fatty acids, all unsaturated (oleic) fatty acids, or any combination of oleic and palmitic acids can be used. Furthermore, positional isomers are possible because the three carbons of glycerol are not equivalent. The triacylglycerols are, in order of increasing melting point: OOO; OOP, OPO; PPO, POP; PPP, where O = oleic and P = palmitic acid. The greater the content of saturated fatty acid, the higher the melting point.

12. **Operational Definition of Lipids** How is the definition of "lipid" different from the definitions of other types of biomolecules that we have considered, such as amino acids, nucleic acids, and proteins?

> **Answer** The word "lipid" does not refer to a particular chemical structure. In contrast to amino acids, nucleic acids, and proteins, lipids are much more chemically diverse. Compounds are categorized as lipids based on their greater solubility in organic solvents than in water.

13. **Effect of Polarity on Solubility** Rank, in order of increasing solubility in water, a triacylglycerol, a diacylglycerol, and a monoacylglycerol, all containing only palmitic acid.

> **Answer** The order of solubility in water is: monacylglycerol > diacylglycerol > triacylglycerol. Increasing the number of palmitic acids increases the hydrophobic portion of the molecule.

14. **Intracellular Messengers from Phosphatidylinositols** When the hormone vasopressin stimulates cleavage of phosphatidylinositol-4,5-bisphosphate by hormone-sensitive phospholipase C, two products are formed. Compare their properties and solubilities in water, and predict whether either would be expected to diffuse readily through the cytosol.

> **Answer** Phosphatidylinositol-4,5-bisphosphate is a membrane lipid. The two products of cleavage are inositol-1,4,5-trisphosphate and diacylglycerol. Diacylglycerol is not water soluble and remains in the membrane, acting as a second messenger. The inositol-1-4, 5-trisphosphate is very water soluble and can diffuse away into the cytosol, also acting as a second messenger.

15. **Identification of Unknown Lipids** Johann Thudichum, who practiced medicine in London about 100 years ago, also dabbled in lipid chemistry in his spare time. He isolated a variety of lipids from neural tissue, and characterized and named many of them. His carefully sealed and labeled vials of isolated lipids were rediscovered many years later. How would you confirm, using techniques available to you but not to him, that the vials he labeled "sphingomyelin" and "cerebroside" actually contain these compounds?

> **Answer** Both sphingomyelin and cerebroside can be chromatographed by silica gel thin-layer- chromatography with known standards, and visualized on the plates by rhodamine. A further test could use acid hydrolysis. Neither compound would be susceptible to mild acid hydrolysis, but harsher conditions would release a free fatty acid in both cases. The fatty acid could be esterified and identified by gas-liquid chromatography.

16. *Analysis of Choline-Containing Phospholipids* How would you distinguish sphingomyelin from phosphatidylcholine by chemical, physical, or enzymatic tests?

> *Answer* Under mild acid hydrolysis, only phosphatidylcholine will release two free fatty acids. Under harsh conditions, phosphatidylcholine will release two fatty acids whereas sphingomyelin will release one. Treatment of both samples with phospholipases A, B, and C will yield different products. Free fatty acids will be released from phosphatidylcholine by lipases A and B, whereas no products will be released from sphingomyelin. With lipase C, phosphatidylcholine will release phosphocholine.

CHAPTER **10** **Biological Membranes and Transport**

1. ***Determining the Cross-Sectional Area of a Lipid Molecule*** When phospholipids are layered gently onto the surface of water, they orient at the air-water interface with their head groups in the water and their hydrophobic tails in the air. The experimental apparatus pictured below **(a)** pushes these lipids together by reducing the surface area available to them. By measuring the force necessary to push the lipids together, it is possible to determine when the molecules are packed tightly together in a continuous monolayer; when that area is approached, the pressure needed to furthur reduce the surface area increases sharply. **(b)** How would you use such an experimental apparatus to determine the average area occupied by a single lipid molecule in a lipid monolayer?

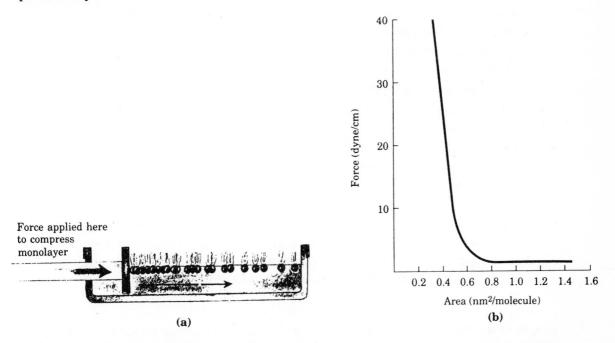

(a) (b)

Answer Determine the surface area of the water at which the pressure increases sharply. Divide this surface area by the number of lipid molecules on the surface, which is calculated by multiplying the number of moles (calculated from the concentration and the molecular weight) by Avogadro's number.

2.   *Evidence for Lipid Bilayer*   In 1925, E. Gorter and F. Grendel used an apparatus like that described in Problem 1 to determine the surface area of a lipid monolayer formed by lipids extracted from erythrocytes of several animal species. They used a microscope to measure the dimensions of individual cells, from which they calculated the average surface area of one erythrocyte. They obtained the data shown below. Were these investigators justified in concluding that "chromocytes [erythrocytes] are covered by a layer of fatty substances that is two molecules thick" (i.e., a lipid bilayer)?

Animal	Volume of packed cells (mL)	Number of cells (per mm^3)	Total surface area of lipid monolayer from cells (m^2)	Total surface area of one erythrocyte (μm^2)
Dog	40	8,000,000	62	98
Sheep	10	9,900,000	2.95	29.8
Human	1	4,740,000	0.47	99.4

Source: Data from Gorter, E. & Grendel, F (1925) On bimolecular layers of lipids on the chromocytes of the blood, *J. Exp. Med*, **41**, 439-443.

Answer   The conclusions are justified for dog erythrocytes but not for sheep or human erythrocytes. The table provides the total surface area of the lipid monolayer. To determine the monolayer surface area per cell, first calculate the total number of cells. For example for dog erythrocytes, the number of cells is

8×10^6 per mm^3 = 8×10^9 per cm^3 (or per mL).

In 40 mL, there is a total of (40 mL)(8×10^9 cells/mL) = 3.2×10^{11} cells
From the table, this number of cells yielded a monolayer surface area of
62 m^2 = 6.2×10^5 cm^2. Dividing the surface area by the number of cells gives

$$\frac{6.2 \times 10^5 \text{cm}^2}{3.2 \times 10^{11} \text{ cells}} = 1.9 \times 10^{-6} \text{ cm}^2/\text{cell}$$

Comparing this number to the total surface area of one erythrocyte
(98 μm^2 = 0.98×10^{-6} cm^2), we find a 2 to 1 relationship. This result justifies the calculations drawn by the investigators; however, similar calculations for the sheep and human erythrocytes reveal a 1 to 1 relationship. (*Note*: the basis and significance of this is unclear, perhaps resulting from an error in the data in the table.)

3.   *Length of a Fatty Acid Molecule*   The carbon-carbon bond distance for single-bonded carbons such as those in a saturated fatty acyl chain is about 0.15 nm. Estimate the length of a single molecule of palmitic acid in its fully extended form. If two molecules of palmitic acid were laid end to end, how would their total length compare with the thickness of the lipid bilayer in a biological membrane?

Answer   Given that the C—C bond length is 0.15nm and that the bond angle of tetrahedral carbon is 109°, the distance between the first and third carbons in an acyl chain is about 0.26 nm. Thus the distance between two adjacent carbons is about 0.13 nm. For palmitic acid, the length is
16 x 0.13 nm ≈ 2.1 nm. Thus two palmitic acids end to end (as in a bilayer) would extend 4.2 nm. This is slightly more than the thickness of a lipid bilayer (4 nm).

4. *Temperature Dependence of Lateral Diffusion* The experiment described in Figure 10-12 was done at 37 °C. If, instead, the whole experiment were carried out at 10 °C, what effect would you predict on the rate of cell-cell fusion, and the rate of membrane protein mixing? Why?

    Answer When the temperature drops, the fluidity of a membrane decreases. This is caused by a decrease in the rate of diffusion of lipids and of the proteins associated with the lipids. Consequently, all processes depending on diffusion, such as cell-cell fusion and protein mixing, would slow down.

5. *Synthesis of Gastric Juice: Energetics* Gastric juice (pH 1.5) is produced by pumping HCl from blood plasma (pH 7.4) into the stomach. Calculate the amount of free energy required to concentrate the H^+ in 1 L of gastric juice at 37 °C. Under cellular conditions, how many moles of ATP must be hydrolyzed to provide this amount of free energy? (The free-energy change for ATP hydrolysis under cellular conditions is about -58 kJ/mol, as we will explain in Chapter 13.)

    Answer Given that pH = $-\log[H^+]$, then $[H^+] = 10^{-pH}$ = -antilog pH
    At pH 1.5, $[H^+] = 10^{-1.5} = 3.16 \times 10^{-2}$ M
    At pH 7.4, $[H^+] = 10^{-7.4} = 3.98 \times 10^{-8}$ M
    $\Delta G_t = RT \ln (C_2/C_1)$
    $\qquad = (8.315 \text{ J/mol·k}) (310 \text{ k}) \ln \left(\dfrac{3.16 \times 10^{-2} \text{ M}}{3.98 \times 10^{-8} \text{ M}} \right) = 35$ kJ/mol

    The amount of ATP required to provide 35 kJ is

    $\dfrac{35 \text{ kJ}}{58 \text{ kJ/mol}} = 0.6$ mol

6. *Energetics of the $Na^+ K^+$ ATPase* The concentration of Na^+ inside a vertebrate cell is about 12 mM, and the cell is bathed in blood plasma containing about 145 mM Na^+. For a typical cell with a transmembrane potential of -0.07 V (inside negative relative to outside), what is the free-energy change for transporting 1 mol of Na^+ out of the cell at 37 °C?

    Answer
    $\Delta G_t = RT \ln (C_2/C_1) + ZF\Delta\Psi$
    $\qquad = (8.315 \text{ J/mol·k})(310\text{k}) \ln \left(\dfrac{145 \text{ mM}}{12 \text{ mM}} \right) + (1)(96,480 \text{ J/V·mol})(0.07 \text{ V})$
    $\qquad = 6.42 \text{ kJ/mol} + 6.75 \text{ kJ/mol} \approx 13.2$ kJ/mol

    Note that 6.75 kJ/mol is the membrane potential portion.

7. *Action of Ouabain on Kidney Tissue* Ouabain specifically inhibits the Na^+K^+ ATPase activity of animal tissues but is not known to inhibit any other enzyme. When ouabain is added in graded concentrations to thin slices of living kidney tissue, it inhibits oxygen consumption by 66%. Explain the basis of this observation. What does it tell us about the use of respiratory energy by kidney tissue?

    Answer Oxidative phosphorylation to supply the cell with ATP accounts for the majority of oxygen consumption. A decrease in oxygen consumption by 66% on addition of ouabain indicates that consumption by the Na^+K^+ ATPase accounts for a large percentage of ATP produced, and thus the energy consumed, by kidney tissue.

8. *Membrane Protein Topology* The receptor for the hormone epinephrine in animal cells is an integral membrane protein (M_r 64,000) that is believed to span the membrane seven times. Show that a protein of this size is capable of spanning the membrane seven times. If you were given the amino acid sequence of this protein, how would you go about predicting which regions of the protein form the membrane-spanning helices?

    Answer Assume that the transmembrane portion of the peptide is an α helix. The rise per amino acid (AA) residue of an α-helix is 0.15 nm/AA. The lipid bilayer is approximately 4 nm thick; to span this, (4 nm)/(0.15 nm/AA) = 27 AA are needed. Thus seven spans requires 7 x 27 = 189 residues. A protein of M_r 64,000 (if the average M_r of an amino acid in a peptide is 110) has approximately 64,000/110 = 580 AA residues. To identify potential transmembrane regions, hydropathy plots can be used (see Box 10-2 in the text). The most hydrophobic (hydropathic) stretches are those most likely to pass through the apolar lipid bilayer.

9. *Energetics of Symport* Suppose that you determined experimentally that a cellular transport system for glucose, driven by symport of Na^+, could accumulate glucose to concentrations 25 times greater than in the external medium, while the external [Na^+] was only ten times greater than the intracellular [Na^+]. Is this a violation of the laws of thermodynamics? If not, how do you explain this observation?

    Answer No; the symport may be able to transport more than one equivalent of glucose per Na^+.

10. *Location of a Membrane Protein* An unknown membrane protein, X, can be extracted from disrupted erythrocyte membranes into a concentrated salt solution. Isolated X can be cleaved into fragments by proteolytic enzymes. But treatment of erythrocytes, first with proteolytic enzymes, followed by disruption and extraction of membrane components, yields intact X. In contrast, treatment of erythrocyte "ghosts" (which consist of only membranes, produced by disrupting the cells and washing out the hemoglobin) with proteolytic enzymes, followed by disruption and extraction, yields extensively fragmented X. What do these experiments indicate about the location of X in the plasma membrane? On the basis of this information, do the properties of X resemble those of glycophorin or those of ankyrin?

    Answer Because protein X can be removed by salt treatment, it must be a peripheral protein. Inability to digest the protein with proteases unless the membrane has been disrupted indicates that protein X is located internally, bound to the inner surface of the erythrocyte plasma membrane. This is also the location of ankyrin.

11. *Membrane Self-Sealing* Cell membranes are self-sealing—if they are punctured or disrupted mechanically, they quickly and automatically reseal. What properties of membranes are responsible for this important feature?

    Answer Hydrophobic interactions are the driving force for membrane formation. Because these forces are noncovalent and reversible, after membranes are disrupted they can be readily reannealed.

12. *Lipid Melting Temperatures*   Membrane lipids in tissue samples obtained from different parts of the leg of a reindeer show different fatty acid compositions. Membrane lipids from tissue near the hooves contain a larger proportion of unsaturated fatty acids than lipids from tissue in the upper part of the leg. What is the significance of this observation?

> *Answer*   The temperature of body tissues at the extremities, such as near the hooves, is generally lower than that of tissues closer to the center of the body. To maintain membrane fluidity, as required by the fluid-mosaic model, membranes at lower temperatures must contain a higher percentage of polyunsaturated fatty acids: unsaturated fatty acids lower the melting point of lipid mixtures.

13. *Flip-Flop Diffusion*   The inner face of the human erythrocyte membrane consists predominantly of phosphatidylethanolamine and phosphatidylserine. The outer face consists predominantly of phosphatidylcholine and sphingomyelin. Although the phospholipid components of the membrane can diffuse in the fluid bilayer, this sidedness is preserved at all times. How?

> *Answer*   The energy required to flip a charged polar head group through a hydrophobic lipid bilayer is prohibitively high.

14. *Membrane Permeability*   At pH 7, tryptophan crosses a lipid bilayer membrane about 1,000 times more slowly than does the closely related substance indole [see structure below]. Suggest an explanation for this observation.

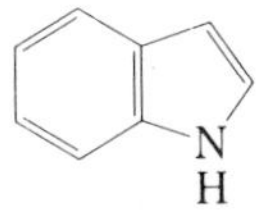

> *Answer*   At pH 7 tryptophan exists as a zwitterion (having both a positive and negative charge), whereas indole is uncharged. The movement of the less polar indole through a hydrophobic membrane is more energetically favorable.

CHAPTER 11   Carbohydrates

1. ***Interconversion of D-Galactose Forms***   A solution of one stereoisomer of a given monosaccharide will rotate plane-polarized light to the left (counterclockwise) and is called the levorotatory isomer, designated (-); the other stereoisomer will rotate plane-polarized light to the same extent but to the right (clockwise) and is called the dextrorotatory isomer, designated (+). An equimolar mixture of the (+) and (-) forms will not rotate plane-polarized light.

The optical activity of a stereoisomer is expressed quantitatively by its *optical rotation*, the number of degrees by which plane-polarized light is rotated on passage through a given path length of a solution of the compound at a given concentration. The *specific rotation* $[\alpha]_D^{25°C}$ of an optically active compound is defined thus:

$$[\alpha]_D^{25°C} = \frac{\text{observed optical rotation (°)}}{\text{length of optical path (dm) x concentration (g/mL)}}$$

The temperature and the wavelength of the light employed (usually the D line of sodium, 589 nm) must be specified in the definition.

A freshly prepared solution of the α form of D-galactose (1 g/mL in a 10 cm cell) shows an optical rotation of +150.7°. When the solution is allowed to stand for a prolonged period of time the observed rotation gradually decreases and reaches an equilibrium value of +80.2°. In contrast, a freshly prepared solution (1 g/mL) of the β form shows an optical rotation of only +52.8°. Moreover, when the solution is allowed to stand for several hours, the rotation increases to an equilibrium value of +80.2°, identical to the equilibrium value reached by α-D-galactose.

(a) Draw the Haworth perspective formulas of the α and β forms of galactose. What feature distinguishes the two forms?

(b) Why does the optical rotation of a freshly prepared solution of the form gradually decrease with time? Why do solutions of the α and β forms (at equal concentrations) reach the same optical rotation at equilibrium?

(c) Calculate the percentage composition of the two forms of galactose at equilibrium.

Answer

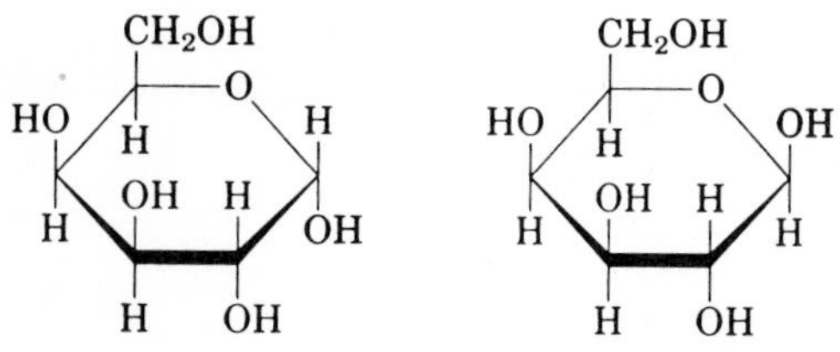

α-ᴅ-Galactose   β-ᴅ-Galactose

(a)   The two forms are distinguished by the configuration of the —OH group on C-1, the anomeric carbon.

(b)   A fresh solution of the α form of galactose will undergo *mutarotation* to an equilibrium mixture containing both the α and β forms. The same applies to a fresh solution of the β form.

(c)   The change in optical rotation from 100% α form (150.7°) to 100% β form (52.8°) is 97.9°. Since an equilibrium mixture rotates light 80.2°, the fraction of galactose in the α form can be calculated as

$$\frac{80.2° - 52.8°}{150.7° - 52.8°} = \frac{27.4°}{97.9°} = 0.279 \approx 28\% \ \alpha \ \text{form, and thus, } 72\% \ \beta \ \text{form.}$$

2.   *Invertase "Inverts" Sucrose*   The hydrolysis of sucrose (specific rotation +66.5°) yields an equimolar mixture of ᴅ-glucose (specific rotation +52.5°) and ᴅ-fructose (specific rotation -92°).

(a)   Suggest a convenient way to determine the rate of hydrolysis of sucrose by an enzyme preparation extracted from the lining of the small intestine.

(b)   Explain why an equimolar mixture of ᴅ-glucose and ᴅ-fructose formed by hydrolysis of sucrose is called invert sugar in the food industry.

(c)   The enzyme invertase (its preferred name is now sucrase) is allowed to act on a solution of sucrose until the optical rotation of the solution becomes zero. What fraction of the sucrose has been hydrolyzed?

Answer

(a)   When completely hydrolyzed, the optical rotation of an equilmolar mixture of ᴅ-glucose and ᴅ-fructose will be 52.5° + -92° = -39.5°. Enzyme (sucrase) activity can be assayed by observing the change in optical rotation starting from a solution of 100% sucrose (α + 66.5°), as it is converted to a 1:1 mixture of ᴅ-glucose and ᴅ-fructose.

(b)   The optical rotation of the mixture is negative (inverted) relative to that of the sucrose solution.

(c)   This is calculated by determining the fraction of maximal optical rotation. When sucrose is completely hydrolyzed, the change in optical rotation (from +66.5° to 39.5°) is 106°. At zero optical rotation (0°), the change in optical rotation is 66.5°. Thus the fraction hydrolyzed = 66.5°/106° = 0.63.

3. *Manufacture of Liquid-Filled Chocolates* The manufacture of chocolates containing a liquid center is an interesting application of enzyme engineering. The flavored liquid center consists largely of an aqueous solution of sugars rich in fructose to provide sweetness. The technical dilemma is the following: the chocolate coating must be prepared by pouring hot melted chocolate over a solid (or almost solid) core, yet the final product must have a liquid, fructose-rich center. Suggest a way to solve this problem. (Hint: the solubility of sucrose is much lower than the solubility of a mixture of glucose and fructose.)

> *Answer* Prepare the core as a semi-solid slurry of sucrose and water. Add a small amount of sucrase, and quickly coat the semi-solid mixture with chocolate. After the chocolate coat has cooled and hardened, the sucrase will hydrolyze enough of the sucrose to reduce the viscosity of the mixture, converting it to a more nearly liquefied mixture of glucose and fructose.

4. *Anomers of Sucrose?* Although lactose exists in two anomeric forms, no anomeric forms of sucrose have been reported. Why?

> *Answer* In contrast to lactose, sucrose has no free anomeric carbon to undergo mutarotation.

5. *Growth Rate of Bamboo* The stems of bamboo, a tropical grass, can grow at the phenomenal rate of 0.3 m/day under optimal conditions. Given that the stems are composed almost entirely of cellulose fibers oriented in the direction of growth, calculate the number of sugar residues per second that must be added enzymatically to growing cellulose chains to account for the growth rate. Each D-glucose unit in the cellulose molecule is about 0.45 nm long.

> *Answer* First calculate the length of growth per second:
>
> $$\frac{0.3 \text{ m/day}}{(24 \text{ h/day})(60 \text{ min/h})(60 \text{ s/min})} = 3.47 \times 10^{-6} \text{ m/s}$$
>
> Since each glucose residue is 0.45 nm (0.45×10^{-9} m) long, the number of residues added per second is
>
> $$\frac{3.47 \times 10^{-6} \text{ m/s}}{0.45 \times 10^{-9} \text{ m/residue}} = 7{,}716 \text{ residues/sec} \approx 7{,}700 \text{ residues/sec}$$

6. *Enzymatic Digestibility of Cellulose and Starch* Both cellulose and α-amylose consist of (1→4)-linked D-glucose units and can be extensively hydrated. Despite this similarity, a person on a diet consisting predominantly of α-amylose (starch) will gain weight, whereas a person on a diet of cellulose (wood) will starve. Why?

> *Answer* Cellulose is a polymer of D-glucose units linked together by (β1→4) linkages, whereas in the amylose polymer the linkages are of the (α1→4) configuration. The enzymes that hydrolyze amylose (amylases) are not active with cellulose. Cellulose ("roughage" in food) is not digested by humans or by most other animals.

7. **Physical Properties of Cellulose and Glycogen**   The practically pure cellulose obtained from the seed threads of the plant genus *Gossypium* (cotton) is tough, fibrous, and completely insoluble in water. In contrast, glycogen obtained from muscle or liver disperses readily in hot water to make a turbid solution. Although they have markedly different physical properties, both substances are composed of (1→4)-linked D-glucose polymers of comparable molecular weight. What features of their structures cause these two polysaccharides to differ in their physical properties? Explain the biological advantages of their respective properties.

> **Answer**   Native cellulose consists of glucose units linked by (β1-4) glycosidic bonds. The β linkages force the polymer chain into an extended conformation (see Fig. 11-17). A parallel series of these extended chains can form intermolecular hydrogen bonds, thus aggregating into long, tough, insoluble fibers. Glycogen consists of glucose units linked by (α1→4) glycosidic bonds. The α linkage causes a bend in the chain and prevents the formation of long fibers. In addition, glycogen is highly branched. Because the conformation of the hydroxyl groups of the glucose units in the polymer exposes them to water, glycogen is highly hydrated and therefore is very water-soluble, so it can be extracted as a dispersion in hot water.
>
> The physical properties of these two polymers are well suited to their biological roles. Cellulose serves as a structural material in plants, consistent with its side-by-side aggregation into tough, insoluble fibers. Glycogen serves as a storage fuel in animals. The highly hydrated glycogen granules with their abundance of free, nonreducing ends can be rapidly hydrolyzed by glycogen phosphorylase to release glucose-1-phosphate.

8. **Glycogen as Energy Storage: How Long Can a Game Bird Fly?**   Since ancient times it had been observed that certain game birds, such as grouse, quail, and pheasants, are easily fatigued. The Greek historian Xenophon wrote: "The bustards, on the other hand, can be caught if one is quick in starting them up, for they will fly only a short distance, like partridges, and soon tire; and their flesh is delicious." The flight muscles of game birds rely almost entirely on the metabolic breakdown of glucose-1-phosphate for the necessary energy, in the form of ATP (see Chapter 14). In game birds, glucose-1-phosphate is formed by the breakdown of stored muscle glycogen, catalyzed by the enzyme glycogen phosphorylase. The rate of ATP production is limited by the rate at which glycogen can be broken down. During a "panic flight," the game bird's rate of glycogen breakdown is quite high, approximately 120 μmol/min of glucose-1-phosphate produced per gram of fresh tissue. Given that the flight muscles usually contain about 0.35% glycogen by weight, calculate how long a game bird can fly.

> **Answer**   The average molecular weight of a glucose unit in glycogen is 160. The amount of usable glucose units (as glycogen) in 1 g of tissue is
>
> $$\frac{0.0035 \text{ g}}{160 \text{ g/mol}} = 2.19 \times 10^{-5} \text{ mol}$$
>
> In 1 min, 120 μmol of glucose-1-phosphate is produced, so 120 μmol of glucose is hydrolyzed. Thus depletion of the glycogen would occur in
>
> $$\frac{2.19 \times 10^{-5} \text{ mol}}{120 \times 10^{-6} \text{ mol/min}} = 0.182 \text{ min} = 10.9 \text{ s} \approx 11 \text{ s}.$$

9. ***Determination of the Extent of Branching in Amylopectin*** The extent of branching (number of ($\alpha1\rightarrow6$) glycosidic bonds) in amylopectin can be determined by the following procedure. A weighted sample of amylopectin is exhaustively treated with a methylating agent (methyl iodide) that replaces all the hydrogens on the sugar hydroxyls with methyl groups, converting —OH to —OCH$_3$. All the glycosidic bonds in the treated sample are then hydrolyzed with aqueous acid. The amount of 2,3-dimethylglucose in the hydrolyzed sample is determined.

2,3-Dimethylglucose

(a) Explain the basis of this procedure for determining the number of ($\alpha1\rightarrow6$) branch points in amylopectin. What happens to the unbranched glucose residues in amylopectin during the methylation and hydrolysis procedure?

(b) A 258 mg sample of amylopectin treated as described above yielded 12.4 mg of 2,3-dimethylglucose. Determine what percentage of the glucose residues in amylopectin contain an ($\alpha1\rightarrow6$) branch.

Answer

(a) The glucose units at branch points are protected from complete methylation because the hydroxyl of C-6 is involved in the branch formation. During methylation and subsequent hydrolysis, the branch points would yield 2,3-dimethylglucose, and the unbranched residues would yield 2,3,6-trimethylglucose.

(b) The average molecular weight of glucose in amylopectin is 160. Thus 258 mg of amylopectin contains

$$\frac{258 \times 10^{-3} \text{ g}}{160 \text{ g/mol}} = 1.61 \times 10^{-3} \text{ mol of glucose}$$

The 12.4 mg yield of 2,3-dimethylglucose (M_r 208) is equivalent to

$$\frac{12.4 \times 10^{-3} \text{ g}}{208 \text{ g/mol}} = 5.96 \times 10^{-5} \text{ mol}$$

Thus the percentage of glucose residues in amylopectin that yield 2,3-dimethylglucose is

$$\frac{5.96 \times 10^{-5} \text{ mol } (100)}{1.61 \times 10^{-3} \text{ mol}} = 3.7\%$$

10. ***Structure Determination of a Polysaccharide*** A polysaccharide of unknown structure was isolated, subjected to exhaustive methylation, and hydrolyzed. Analysis of the products revealed three methylated sugars: 2,3,4-tri-*O*-methyl-D-glucose, 2,4-di-*O*-methyl-D-glucose, and 2,3,4,6-tetra-*O*-methyl-D-glucose, in the ratio 20:1:1. What is the structure of the polysaccharide?

Answer Because the predominant product is 2,3,4 tri-*O*-methyl-D-glucose, the predominant linkage must be 1→6. The formation of 2,4 di-*O*-methyl-D-glucose indicates that the branch point is through C-3. The ratio of methylated sugars indicates that a branch occurs at a frequency of every 20 residues. The 2,3,4,6-tetra-*O*-methyl-D-glucose is derived from nonreducing chain ends, which compose about 1/20 or 5% of the residues, also consistent with a high degree of branching.

11. *Empirical Formula Determination* An unknown substance containing only C, H, and O was isolated from goose liver. A 0.423 g sample produced 0.620 g of CO_2 and 0.254 g of H_2O after complete combustion in excess oxygen. Is the empirical formula of this substance consistent with its being a carbohydrate? Explain.

Answer The stoichiometry of the reaction must first be calculated. The reaction can be written

$$C_?H_?O_? \quad + \quad ?O_2 \quad \longrightarrow \quad CO_2 \quad + \quad H_2O$$

$C_?H_?O_?$	$?O_2$		CO_2		H_2O
(0.423 g)	(?)		(0.620 g)		(0.254 g)
(? mol)	(? mol)		(1.41×10^{-2} mol)		(1.41×10^{-2} mol)

where each ? represents a different unknown value. Because equivalent mole amounts of CO_2 and H_2O are produced, for every C produced, the reaction also produces 2 H and 3 O. The simplest equation to describe this stoichiometry is

$$CH_2O + O_2 \longrightarrow CO_2 + H_2O$$

The empirical formula CH_2O is characteristic of a carbohydrate.

12. *Reaction with Fehling's Reagent* A sample of disaccharide is either lactose or sucrose. No reddish precipitate forms in Fehling's reaction, unless the compound is first warmed in dilute acid. Is it lactose or sucrose? Explain.

Answer It is sucrose. In sucrose the anomeric (reducing) carbons of glucose and fructose are involved in glycosidic linkages. Only after mild acid hydrolysis to glucose and fructose are the anomeric carbons available for reduction of Fe^{3+} to Fe^{2+}, which precipitates as the red oxide. (Note: the answer is incorrectly given as lactose on p. AP-8 of the text.)

13. *Glucose Oxidase in Determination of Blood Glucose* The enzyme glucose oxidase isolated from the mold *Penicillium notatum* catalyzes the oxidation of β-D-glucose to D-glucono-δ-lactone. This enzyme is highly specific for the β anomer of glucose and does not affect the α anomer. In spite of this specificity, the reaction catalyzed by glucose oxidase is commonly used in a clinical assay for total blood glucose—i.e., solutions consisting of a mixture of β- and α-D-glucose. How is this possible? Aside from allowing the detection of smaller quantities of glucose, what advantage does glucose oxidase offer over Fehling's reaction for the determination of blood glucose?

Answer The rate of mutarotation (interconversion of the α and β anomers) is sufficiently high that as the enzyme consumes β-D-glucose, α-D-glucose is further converted to the β form. Eventually all of the glucose is oxidized. Glucose oxidase, which is specific for glucose, will not detect other reducing sugars such as galactose that would react with Fehling's reagent.

14. *Volume of Chondroitin Sulfate in Solution*   One of the critical functions of chondroitin sulfate is to act as a lubricant in skeletal joints by creating a gel-like medium that is resilient to friction and shock. This function appears to be related to a distinctive property of chondroitin sulfate: the volume occupied by the molecule is much greater in solution than in the dehydrated solid. Why is the volume occupied by the molecule so much larger in solution?

> *Answer*   The negative charges of chondroitin sulfate repel each other and force the molecule into an extended conformation. The polar molecule also attracts many water molecules (water of hydration), which further increase the molecular volume.

15. *Information Content of Oligosaccharides*   The carbohydrate portion of some glycoproteins may serve as a cellular recognition site. In order to perform this function, the oligosaccharide moiety of glycoproteins must have the potential to occur in a large variety of forms. Which can produce a larger variety of structures: oligopeptides composed of five different amino acid residues or oligosaccharides composed of five different monosaccharide residues? Explain.

> *Answer*   Oligosaccharides; their subunits can be combined in more ways than the amino acids subunits of oligopeptides. Each of the several hydroxyl groups of each monosaccharide can participate in glycodisic bonds, and the configuration of each glycosidic bond may be either α or β. Furthermore, the polymer can be linear or branched. Oligopeptides are unbranched polymers, with all amino acid units linked through identical peptide bonds.

CHAPTER **12** **Nucleotides and Nucleic Acids**

1. *Determination of Protein Concentration by UV Absorption in a Solution Containing Nucleic Acids* The concentration of protein or nucleic acid in solutions containing both can be estimated by using their light absorption properties. Proteins have a strong absorption centered at a wavelength of 280 nm, whereas nucleic acids absorb most strongly at 260 nm. When both proteins and nucleic acids are present in a solution, their respective concentrations can be estimated by measuring the absorbance (A) of the solution at 280 nm and 260 nm and using the table below. $R_{280/260}$ is the ratio of the absorbance at 280 and 260 nm. The table indicates the percentage of total mass that is nucleic acid, and provides a factor, F, to correct the A_{280} reading and give a more accurate protein estimate. The protein concentration (in mg/mL) is equal to F x A_{280} (assuming the cuvette is 1 cm wide). What are the protein and nucleic acid concentration if $A_{280} = 0.69$ and $A_{260} = 0.94$?

$R_{280/260}$	(%)	F	$R_{280/260}$	(%)	F
1.75	0.00	1.116	0.846	5.50	0.656
1.63	0.25	1.081	0.822	6.00	0.632
1.52	0.50	1.054	0.804	6.50	0.607
1.40	0.75	1.023	0.784	7.00	0.585
1.36	1.00	0.994	0.767	7.50	0.565
1.30	1.25	0.970	0.753	8.00	0.545
1.25	1.50	0.944	0.730	9.00	0.508
1.16	2.00	0.899	0.705	10.00	0.478
1.09	2.50	0.852	0.671	12.00	0.422
1.03	3.00	0.814	0.644	14.00	0.377
0.979	3.50	0.776	0.615	17.00	0.322
0.939	4.00	0.743	0.595	20.00	0.278
0.874	5.00	0.682			

Answer The ratio of A_{280} to A_{260} is

$R_{280/260} = 0.69/0.94 = 0.73$

From the table, the F value for this $R_{280/260}$ is 0.508.
The protein concentration can be calculated by multiplying A_{280} by F:

$(0.69)(0.508) = 0.35$ mg/mL

Again from the table, for $R_{280/260} = 0.73$, nucleic acid accounts for 9% of the total mass; thus protein accounts for 91%. The total mass/mL is (0.35 mg/mL)/0.91 = 0.385 mg/mL. The concentration of nucleic acid = (0.385 - 0.35) mg/mL = 0.035 mg/mL.

2. *Nucleotide Structure*   What positions in a purine ring have the potential to form hydrogen bonds, but are not involved in the hydrogen bonds of Watson-Crick base pairs?

   Answer All the ring nitrogens (N-1, N-3, N-7, and N-9) have the potential to form hydrogen bonds (see Figs. 12-1, 12-11 and 4-3). However, N-1 is involved in hydrogen bonding when a purine is base paired with a pyrimidine in a Watson-Crick base pair. Thus, the answer is N-3, N-7, and N-9.

3. *Base Sequence of Complementary DNA Strands*   Write the base sequence of the complementary strand of double-helical DNA in which one strand has the sequence (5')ATGCCCGTATGCATTC(3').

   Answer The two strands of a DNA molecule are antiparallel and their sequences are complementary: (3')TACGGGCATACGTAAG (5').

4. *DNA of the Human Body*   Calculate the weight in grams of a double-helical DNA molecule stretching from the earth to the moon ($\sim$ 320,000 km). The DNA double helix weighs about 1×10^{-18} g per 1,000 nucleotide pairs; each base pair extends 0.34 nm. For an interesting comparison, your body contains about 0.5 g of DNA!

   Answer The length of the DNA is:
   $(320,000 \text{ km})(10^{12} \text{ nm/km}) = 3.2 \times 10^{17} \text{ nm}$.
   The number of base pairs (bp) is
   $$\frac{3.2 \times 10^{17} \text{ nm}}{0.34 \text{ nm/bp}} = 9.4 \times 10^{17} \text{ bp}$$
   Thus, the weight of the DNA molecule is
   $(9.4 \times 10^{17} \text{ bp})(1 \times 10^{-18} \text{ g}/10^3 \text{ bp}) = 9.4 \times 10^{4} \text{ g} = 0.94 \times 10^{3} \text{ g}$

5. *DNA Bending*   Assume that a poly(A) tract five base pairs long produces a bend of about 20°. Calculate the total (net) bend produced in the DNA if the center base pairs (the third of five) of two successive (dA)₅ tracts are located (a) 10 or (b) 15 base pairs apart. Assume that there are 10 base pairs per turn in the DNA double helix.

   Answer When bending elements are repeated in phase with the helix screw (that is, every 10 base pairs) as in (a), the total bend is additive; when bending elements are separated by a half-integral number of turns as in (b), the bends are out of phase and thus cancel each other out. Thus the net bend is (a) 40°; (b) 0°.

6. *Distinction between DNA Structure and RNA Structure*   Hairpins may form at palindromic sequences in single strands of either RNA or DNA. How is the helical structure of a hairpin in RNA different from that of a hairpin in DNA?

   Answer The RNA helix assumes the A conformation; the DNA helix generally assumes the B conformation. The presence of 2'—OH group on ribose makes it sterically impossible for double helical RNA to assume the B-form helix.

7. ***Nucleotide Chemistry***   In the cells of many eukaryotic organisms, there are highly specialized systems that specifically repair G-T mismatches in DNA. The mismatch is repaired to form a $G \equiv C$ base pair (not $A = T$). This G-T mismatch repair system occurs in addition to a more general system that repairs virtually all mismatches. Can you think of a reason why cells require a specialized system to repair G-T mismatches?

> ***Answer***   The C residue of a CpG sequence in eukaryotic DNA is often methylated at the 5′ position to yield 5-methylcytosine. Spontaneous deamination of 5-methylcytosine yields thymine, T. A G-T mismatch resulting from spontaneous deamination of 5-methylcytosine in a $G \equiv C$ base pair is one of the most common mismatches in eukaryotic cells. A specialized repair mechanism is required to repair these mismatches.

8. ***Nucleic Acid Structure***   Explain why there is an increase in the absorption of UV light (hyperchromic effect) when double-stranded DNA is denatured.

> ***Answer***   The double-helical structure is stabilized by hydrogen bonding between complementary bases on opposite strands, and base stacking between adjacent bases on the same strand. Base stacking in nucleic acids causes a decrease in the absorption of UV light. Denaturation of DNA results in loss of base stacking, with an increase in UV absorption.

9. ***Base Pairing in DNA***   In samples of DNA isolated from two unidentified species of bacteria, adenine makes up 32 and 17%, respectively, of the total bases. What relative proportions of adenine, guanine, thymine, and cytosine would you expect to find in the two DNA samples? What assumptions have you made? One of these bacteria was isolated from a hot spring (64 °C). Which DNA came from this thermophilic bacterium? What is the basis for your answer?

> ***Answer***   For any double-helical DNA, A = T and G = C. Since the $G \equiv C$ base pair involves three hydrogen bonds and the $A = T$ base pair involves two hydrogen bonds, the higher the G + C content of a DNA molecule, the higher the melting temperature. The DNA sample containing 32% A must contain 32% T, 18% G, and 18% C. The DNA sample containing 17% A must contain 17% T, 33% G, and 33% C. This calculation is based on the assumption that both DNA molecules are double-stranded. The second DNA sample, with the higher G + C content (66%), came from the thermophilic bacterium, since this DNA has a higher melting temperature and thus is more stable at high temperatures.

CHAPTER 13

Principles of Bioenergetics

1. ***Entropy Changes during Egg Development*** Consider an ecosystem consisting of an egg in an incubator. The white and yolk of the egg contain proteins, carbohydrates, and lipids. If fertilized, the egg is transformed from a single cell to a complex organism. Discuss this irreversible process in terms of the entropy changes in the system, surroundings, and universe. Be sure that you first clearly define the system and surroundings.

   Answer Consider the developing chick as the system. The nutrients, egg shell, and outside world are the surroundings. Transformation of the single cell into a chick drastically reduces the entropy of the system (increases the order). Initially, the parts of the egg outside the embryo (within the surroundings) contain complex fuel molecules (low entropy). During incubation, some of these complex molecules are converted into large numbers of CO_2 and H_2O molecules (high entropy). This increase in entropy of the surroundings is larger than the decrease in entropy of the chick (the system). Thus the entropy of the universe (the system + surroundings) increases.

2. ***Calculation of $\Delta G^{\circ\prime}$ from Equilibrium Constants*** Calculate the standard free-energy changes of the following metabolically important enzyme-catalyzed reactions at 25 °C and pH 7.0 from the equilibrium constants given.

   (a) Glutamate + oxaloacetate $\xrightleftharpoons[\text{aminotransferase}]{\text{aspartate}}$ aspartate + α-ketoglutarate $K'_{eq} = 6.8$

   (b) Dihydroxyacetone phosphate $\xrightleftharpoons[\text{isomerase}]{\text{triose phosphate}}$ glyceraldehyde-3-phosphate $K'_{eq} = 0.0475$

   (c) Fructose-6-phosphate + ATP $\xrightleftharpoons{\text{phosphofructokinase}}$

   fructose-1,6-bisphosphate + ADP $K'_{eq} = 254$

   Answer
   $\Delta G = \Delta G^{\circ\prime} + RT \ln K'_{eq}$
   At equilibrium $\Delta G = 0$ and $\Delta G^{\circ\prime} = -RT \ln K'_{eq}$
   where $R = 8.315$ J/ mol·K and
   $T = 25$ °C $= 298$ K, $RT = 2.479$ kJ/mol. We can now calculate the $\Delta G^{\circ\prime}$ values from the K'_{eq} for each reaction.

   (a) $\Delta G^{\circ\prime} = -(2.479 \text{ kJ/mol}) \ln 6.8 = -4.75 \text{ kJ/mol}$

   (b) $\Delta G^{\circ\prime} = -(2.479 \text{ kJ/mol}) \ln 0.0475 = 7.6 \text{ kJ/mol}$

   (c) $\Delta G^{\circ\prime} = -(2.479 \text{ kJ/mol}) \ln 254 = -13.7 \text{ kJ/mol}$

3. *Calculation of Equilibrium Constants from* **ΔG°′**   Calculate the equilibrium constants K'_{eq} for each of the following reactions at pH 7.0 and 25 °C, using the ΔG°′ values of Table 13-4:

(a) Glucose-6-phosphate + H_2O $\underset{\text{6-phosphatase}}{\overset{\text{glucose-}}{\rightleftharpoons}}$ glucose + P_i

(b) Lactose + H_2O $\overset{\text{β-galactosidase}}{\rightleftharpoons}$ glucose + galactose

(c) Malate $\overset{\text{fumarase}}{\rightleftharpoons}$ fumarate + H_2O

Answer

$\Delta G = \Delta G°′ + RT \ln K'_{eq}$

At equilibrium $\Delta G = 0$, so $\ln K'_{eq} = -\Delta G°′/RT$

or $K'_{eq} = e^{-(\Delta Go′/RT)}$ where RT = 2.479 kJ/mol at 25 °C (See Problem 2).

From these relationships, we can calculate K'_{eq} for each reaction using the values given for ΔG°′ in Table 13-4.

(a) For G-6-phosphatase:

$\Delta G°′$ = -13.8 kJ/mol

$\ln K'_{eq}$ = (-13.8 kJ/mol)/(2.479 kJ/mol)

          = 5.567

K'_{eq} = 262 M

(b) For β-galactosidase:

$\Delta G°′$ = -15.9 kJ/mol

$\ln K'_{eq}$ = (-15.9 kJ/mol)/(2.479 kJ/mol)

          = 6.414

K'_{eq} = 610 M

(c) For fumarase:

$\Delta G°′$ = 3.1 kJ/mol

$\ln K'_{eq}$ = -(3.1 kJ/mol)/(2.479 kJ/mol)

          = -1.251

K'_{eq} = 0.29

4. *Experimental Determination of* **K′$_{eq}$** *and* **ΔG°′**   If a 0.1 M solution of glucose-1-phosphate is incubated with a catalytic amount of phosphoglucomutase, the glucose-1-phosphate is transformed to glucose-6-phosphate until equilibrium is established. The equilibrium concentrations are

Glucose-1-phosphate $\rightleftharpoons$ Glucose-6-phosphate
4.5×10^{-3} M 9.6×10^{-2} M

Calculate K'_{eq} and ΔG°′ for this reaction at 25 °C.

Answer

K'_{eq} = [G6P]/[G1P] = $(9.6 \times 10^{-2}$ M$)/(4.5 \times 10^{-3}$ M$)$

        = 21.3 ≈ 21

$\Delta G°′$ = $-RT \ln K'_{eq}$

        = -(2.479 kJ/mol)(ln 21.3) = -7.6 kJ/mol

5. *Experimental Determination of ΔG°′ for ATP Hydrolysis* A direct measurement of the standard free-energy change associated with the hydrolysis of ATP is technically demanding because the minute amount of ATP remaining at equilibrium is difficult to measure accurately. The value of ΔG°′ can be calculated indirectly, however, from the equilibrium constants of two other enzymatic reactions having less favorable equilibrium constants:

Glucose-6-phosphate + H$_2$0 $\longrightarrow$ glucose + P$_i$ K'_{eq} = 270

ATP + glucose $\longrightarrow$ ATP + glucose-6-phosphate K'_{eq} = 890

Using this information, calculate the standard free energy of hydrolysis of ATP. Assume a temperature of 25 °C.

Answer The reactions, if coupled together, constitute a "futile cycle" that results in the net hydrolysis of ATP:
(1) G6P + H$_2$O $\longrightarrow$ glucose + P$_i$
(2) ATP + glucose $\longrightarrow$ G6P + ADP
_____

Sum: ATP + H$_2$O $\longrightarrow$ ADP + P$_i$

$\Delta G°'$ = -$RT \ln K'_{eq}$
$\Delta G_1°'$ = (-2.479 kJ/mol)(ln 270) = -13.9 kJ/mol
$\Delta G_2°'$ = (-2.479 kJ/mol)(ln 890) = -16.8 kJ/mol
$\Delta G_{sum}°'$ = $\Delta G_1°'$ + $\Delta G_2°'$ = -30.7 kJ/mol

6. *Difference between ΔG°′ and ΔG* Consider the following interconversion, which occurs in glycolysis (Chapter 14).

Fructose-6-phosphate $\rightleftharpoons$ glucose-6-phosphate K'_{eq} = 1.97

(a) What is ΔG°′ for the reaction (assuming that the temperature is 25 °C)?

(b) If the concentration of fructose-6-phosphate is adjusted to 1.5 M and that of glucose-6-phosphate is adjusted to 0.5 M, what is ΔG?

(c) Why are ΔG°′ and ΔG different?

Answer
(a) At equilibrium, $\Delta G°'$ = -$RT \ln K'_{eq}$
= -(2.479 kJ/mol) ln 1.97
= -1.68 kJ/mol ≈ -1.7 kJ/mol

(b) By definition, ΔG = $\Delta G°'$ + $RT \ln K'_{eq}$
K'_{eq} = [G6P]/[F6P] = 0.5 M/1.5 M = 0.33
ΔG = -1.68 kJ/mol + (2.479 kJ/mol) ln 0.33
= -4.4 kJ/mol

(c) ΔG°′ for any reaction is a fixed parameter, since it is defined for standard conditions of temperature (25 °C = 298 K) and concentration (both F6P and G6P at 1 M). In contrast, ΔG is a variable and can be calculated for any set of product and reactant concentrations. ΔG is defined as ΔG°′ (standard conditions) plus whatever differences occur in this parameter upon moving to nonstandard conditions.

7. *Dependence of ΔG on pH*   The free energy released by the hydrolysis of ATP under standard conditions at pH 7.0 is -30.5 kJ/mol. If ATP is hydrolyzed under standard conditions but at pH 5.0, is more or less free energy released? Why?

    Answer Less; the overall equation for ATP hydrolysis can be approximated as:

    $$ATP^{4-} + H_2O \rightleftharpoons ADP^{3-} + HPO_4^{2-} + H^+$$

    Because H^+ ions are produced in the reaction, if the pH at which this reaction is carried out is lower—that is, as $[H^+]$ increases—the equilibrium is shifted to the left (toward the reactants).   As a result, at lower pH values the reaction does not proceed as far toward products, and less free energy is released.

8. *The ΔG°′ for Coupled Reactions*   Glucose-1-phosphate is converted into fructose-6-phosphate in two successive reactions:

    Glucose-1-phosphate $\longrightarrow$ glucose-6-phosphate
    Glucose-6-phosphate $\longrightarrow$ fructose-6-phosphate

    Using the $\Delta G°′$ values in Table 13-4, calculate the equilibrium constant, K'_{eq}, for the sum of the two reactions at 25 °C.

    Answer

    | (1) | G1P $\longrightarrow$ G6P | $\Delta G_1°′$ | = -7.3 kJ/mol |
    | (2) | G6P $\longrightarrow$ F6P | $\Delta G_2°′$ | = 1.7 kJ/mol |
    | | | | |
    | Sum: | G1P $\longrightarrow$ F6P | $\Delta G_{sum}°′$ | = -5.6 kJ/mol |

    $\ln K'_{eq} = -\Delta G°′/RT$
    $\qquad = -(-5.6 \text{ kJ/mol})/(2.479 \text{ kJ/mol})$
    $\qquad = 2.259$
    $K'_{eq} \quad = 9.57 \approx 9.6$

9. *Strategy for Overcoming an Unfavorable Reaction: ATP-Dependent Chemical Coupling*
    The phosphorylation of glucose to glucose-6-phosphate is the initial step in the catabolism of glucose. The direct phosphorylation of glucose by P_i is described by the equation

    Glucose + P_i $\longrightarrow$ glucose-6-phosphate + H_2O $\Delta G°′ = 13.8$ kJ/mol

    (a) Calculate the equilibrium constant for the above reaction. In the rat hepatocyte the physiological concentrations of glucose and P_i are maintained at approximately 4.8 mM. What is the equilibrium concentration of glucose-6-phosphate obtained by the direct phosphorylation of glucose by P_i? Does this route represent a reasonable metabolic route for the catabolism of glucose? Explain.

    Answer
    $\Delta G°′ \quad = -RT \ln K'_{eq}$
    $\ln K'_{eq} = -\Delta G°′/RT$
    $\qquad = -(13.8 \text{ kJ/mol})/(2.479 \text{ kJ/mol})$
    $K'_{eq} \quad = e^{-5.567}$
    $\qquad = 3.8 \times 10^{-3} \text{ M}^{-1}$

Note: this value has units M^{-1} because the expression for K'_{eq} from the chemical equilibrium includes H_2O—see below.)

$$K'_{eq} \quad = \quad \frac{[G6P]}{[Glc][P_i]}$$

$$
\begin{aligned}
[G6P] &= K'_{eq}[Glc][P_i] \\
&= (3.8 \times 10^{-3}\ M^{-1})(4.8 \times 10^{-3}\ M)(4.8 \times 10^{-3}\ M) \\
&= 8.8 \times 10^{-8}\ M
\end{aligned}
$$

This would not be a reasonable route for glucose catabolism because the cellular [G6P] is likely be much higher than 8.8×10^{-8} M, and the reaction would be unfavorable.

(b) In principle, at least, one way to increase the concentration of glucose-6-phosphate is to drive the equilibrium reaction to the right by increasing the intracellular concentrations of glucose and P_i. Assuming a fixed concentration of P_i at 4.8 mM, how high would the intracellular concentration of glucose have to be to have an equilibrium concentration of glucose-6-phosphate of 250 μM (normal physiological concentration)? Would this route be a physiologically reasonable approach, given that the maximum solubility of glucose is less than 1 M?

Answer Because $K'_{eq} \quad = \quad \dfrac{[G6P]}{[Glc][P_i]}$

then $[Glc] \quad = \quad \dfrac{[G6P]}{K'_{eq}\ [P_i]}$

$$= \quad \frac{250 \times 10^{-6}\ M}{(3.8 \times 10^{-3}\ M^{-1})(4.8 \times 10^{-3}\ M)} \quad = \quad 14\ M$$

This would not be a reasonable route because the maximum solubility of glucose is less than 1 M.

(c) The phosphorylation of glucose in the cell is coupled to the hydrolysis of ATP; that is, part of the free energy of ATP hydrolysis is utilized to effect the endergonic phosphorylation of glucose:

(1)	Glucose + P_i $\longrightarrow$ glucose-6-phosphate + H_2O	$\Delta G^{\circ\prime} = 13.8$ kJ/mol
(2)	ATP + H_2O $\longrightarrow$ ADP + P_i	$\Delta G^{\circ\prime} = -30.5$ kJ/mol

Sum: Glucose + ATP $\longrightarrow$ glucose-6-phosphate + ADP

Calculate K'_{eq} for the overall reaction. When the ATP-dependent phosphorylation of glucose is carried out, what concentration of glucose is needed to achieve a 250 μM intracellular concentration of glucose-6-phosphate when the concentrations of ATP and ADP are 3.38 and 1.32 mM, respectively? Does this coupling process provide a feasible route, at least in principle, for the phosphorylation of glucose as it occurs in the cell? Explain.

Answer

(1)	Glc + P_i $\longrightarrow$ G6P + H_2O	$\Delta G^{\circ}_1{}'$	$= 13.8$ kJ/mol
(2)	ATP + H_2O $\longrightarrow$ ADP + P_i	$\Delta G^{\circ}_2{}'$	$= -30.5$ kJ/mol

Sum: Glc + ATP $\longrightarrow$ G6P + ADP $\Delta G_{sum}{}^{\circ\prime}$ $= -16.7$ kJ/mol

$$\ln K'_{eq} = -\Delta G^{\circ\prime}/RT$$
$$= -(-16.7 \text{ kJ/mol})/(2.479 \text{ kJ/mol})$$
$$= 6.74$$

$$K'_{eq} = 843$$

Since, $K'_{eq} = \dfrac{[G6P][ADP]}{[Glc][ATP]}$

then $[Glc] = \dfrac{[G6P][ADP]}{(K'_{eq})[ATP]}$

$$= \dfrac{(250 \times 10^{-6} \text{ M})(1.32 \times 10^{-3} \text{ M}^{-1})}{(843)(3.38 \times 10^{-3} \text{ M})}$$

$$= 1.2 \times 10^{-7} \text{ M}$$

This route is feasible because the [Glc] is reasonable.

(d) Although coupling ATP hydrolysis to glucose phosphorylation makes thermodynamic sense, how this coupling is to take place has not been specified. Given that coupling requires a common intermediate, one conceivable route is to use ATP hydrolysis to raise the intracellular concentration of P_i and thus drive the unfavorable phosphorylation of glucose by P_i. Is this a reasonable route? Explain.

Answer No; this is not reasonable. Such a high P_i concentration would be required when glucose is at its physiological level that phosphate salts of divalent cations would precipitate out.

(e) The ATP-coupled phosphorylation of glucose is catalyzed in the hepatocyte by the enzyme glucokinase. This enzyme binds ATP and glucose to form a glucose-ATP-enzyme complex, and the phosphate is transferred directly from ATP to glucose. Explain the advantages of this route.

Answer Direct transfer of the phosphate group from ATP to glucose takes advantage of the high group transfer potential of ATP and does not demand that the concentration of intermediates be very high, unlike the mechanism proposed in (d). In addition, the usual benefits of enzymatic catalysis apply, including binding interactions between the enzyme and its substrates; induced fit leading to the exclusion of water from the active site, so that only glucose is phosphorylated; and stabilization of the transition state.

10. *Calculations of $\Delta G^{\circ\prime}$ for ATP-Coupled Reactions* From data in Table 13-6 calculate $\Delta G^{\circ\prime}$ value for the reactions

(a) Phosphocreatine + ADP $\longrightarrow$ creatine + ATP

(b) ATP + fructose $\longrightarrow$ ADP + fructose-6-phosphate

Answer

(a) The $\Delta G^{\circ\prime}$ value for the overall reaction is calculated from the sum of the $\Delta G^{\circ\prime}$ values for the coupled reactions.

(1) Phosphocreatine + $H_2O \longrightarrow$ creatine + P_i	$\Delta G^{\circ}_1{}'$	= -43.0 kJ/mol
(2) ADP + $P_i \longrightarrow$ ATP + H_2O	$\Delta G^{\circ}_2{}'$	= 30.5 kJ/mol
Sum: Phosphocreatine + ADP $\longrightarrow$ creatine + ATP	$\Delta G_{sum}{}^{\circ\prime}$	= -12.5 kJ/mol

(b)

(1)	ATP + H_2O $\longrightarrow$ ADP + P_i	$\Delta G^{\circ}_1{}'$	= -30.5 kJ/mol
(2)	Fructose + P_i $\longrightarrow$ F6P + H_2O	$\Delta G^{\circ}_2{}'$	= 15.9 kJ/mol

Sum: ATP + fructose $\longrightarrow$ ADP + F6P	$\Delta G_{sum}{}^{\circ}{}'$	= -14.6 kJ/mol

11. *Coupling ATP Cleavage to an Unfavorable Reaction* This problem explores the consequences of coupling ATP hydrolysis under physiological conditions to a thermodynamically unfavorable biochemical reaction. Because we want to explore these consequences in stages, we shall consider the hypothetical transformation, X $\longrightarrow$ Y, a reaction for which $\Delta G^{\circ}{}' = 20$ kJ/mol.

(a) What is the ratio [Y]/[X] at equilibrium?

(b) Suppose X and Y participate in a series of reactions during which ATP is hydrolyzed to ADP and P_i. The overall reaction is

$$X + ATP + H_2O \longrightarrow Y + ADP + P_i$$

Calculate [Y]/[X] for this reaction at equilibrium. Assume for the purposes of this calculation that the concentrations of ATP, ADP, and P_i are all 1 M when the reaction is at equilibrium.

(c) We know that [ATP], [ADP], and [P_i] are *not* 1 M under physiological conditions. Calculate the ratio [Y]/[X] for the ATP-coupled reaction when the values of [ATP], [ADP], and [P_i] are those found in rat myocytes (Table 13-5).

Answer

(a) The ratio $[Y]_{eq}/[X]_{eq}$ is equal to the equilibrium constant, K'_{eq}.

$$\ln K'_{eq} = -\Delta G^{\circ}{}'/RT$$
$$= -(20 \text{ kJ/mol})/(2.479 \text{ kJ/mol})$$
$$= -8.07$$
$$K'_{eq} = e^{-8.07} = 3.02 \times 10^{-4} = [Y]_{eq}/[X]_{eq}$$

This is a very small value of K'_{eq}; consequently, $\Delta G^{\circ}{}'$ is large and positive, making the reaction energetically unfavorable as written.

(b) First we need to calculate $\Delta G^{\circ}{}'$ for the overall reaction.

(1)	X $\longrightarrow$ Y	$\Delta G^{\circ}_1{}'$	= 20 kJ/mol
(2)	ATP + H_2O $\longrightarrow$ ADP + P_i	$\Delta G^{\circ}_2{}'$	= -30.5 kJ/mol

Sum: X + ATP $\longrightarrow$ ADP + P_i + Y	$\Delta G_{sum}{}^{\circ}{}'$	= -10.5 kJ/mol

$$K'_{eq} = \frac{[Y]_{eq} [P_i]_{eq} [ADP]_{eq}}{[X]_{eq} [ATP]_{eq}} \quad ; \quad \text{note: water is omitted.}$$

Since [ADP], [ATP], and [P_i] are 1 M, this simplifies to $K'_{eq} = [Y]/[X]$ in units of M.

$$\ln K'_{eq} = -\Delta G^{\circ}{}'/RT$$
$$= -(-10.5 \text{ kJ/mol})/(2.479 \text{ kJ/mol}) = 4.236$$
$$K'_{eq} = e^{4.236} = 69.1 = [Y]/[X]$$

$\Delta G^{\circ}{}'$ is fairly large and negative; the coupled reaction is favorable as written.

(c) Here we are dealing with the nonstandard conditions of the cell. Under physiological conditions, a favorable reaction (under standard conditions) becomes even more favorable.

$$K'_{eq} = \frac{[Y]_{eq}\,[P_i]_{eq}\,[ADP]_{eq}}{[X]_{eq}\,[ATP]_{eq}}$$

$$[Y]/[X] = \frac{K'_{eq}\,[ATP]}{[P_i][ADP]}$$

$$= \frac{(69.1\ M)(8.05 \times 10^{-3}\ M)}{(8.05 \times 10^{-3}\ M)(0.93 \times 10^{-3}\ M)}$$

$$= 7.4 \times 10^4$$

12. *Calculations of* Δ*G at Physiological Concentrations*   Calculate the physiological ΔG (not $\Delta G^{o\prime}$) for the reaction

Phosphocreatine + ADP $\longrightarrow$ creatine + ATP

at 25 °C as it occurs in the cytosol of neurons, in which phosphocreatine is present at 4.7 mM, creatine at 1.0 mM, ADP at 0.20 mM, and ATP at 2.6 mM.

Answer   Using $\Delta G^{o\prime}$ values from Table 13-6:

(1) Phosphocreatine + H_2O $\longrightarrow$ creatine + P_i	$\Delta G_1^{o\prime}$	= -43.0 kJ/mol
(2) ADP + P_i $\longrightarrow$ ATP + H_2O	$\Delta G_2^{o\prime}$	= 30.5 kJ/mol

Sum: Phosphocreatine + ADP $\longrightarrow$ creatine + ATP $\Delta G_{sum}^{o\prime}$ = -12.5 kJ/mol

$$K'_{eq} = \frac{[creatine][ATP]}{[phosphocreatine][ADP]}$$
$$= \frac{(1 \times 10^{-3}\ M)(2.6 \times 10^{-3}\ M)}{(4.7 \times 10^{-3}\ M)(2 \times 10^{-4}\ M)}$$
$$= 2.76$$
$$\Delta G = \Delta G^{o\prime} + RT \ln K'_{eq}$$
$$= -12.5\ kJ/mol + (2.479\ kJ/mol) \ln 2.76$$
$$= -9.98\ kJ\ mol \approx -10\ kJ/mol$$

13. *Free Energy Required for ATP Synthesis under Physiological Conditions*   In the cytosol of rat hepatocytes, the mass-action ratio is

$$\frac{[ATP]}{[ADP][P_i]} = 5.33 \times 10^2\ M^{-1}$$

Calculate the free energy required to synthesize ATP in the rat hepatocyte.

Answer   The reaction for the synthesis of ATP is:

ADP + P_i $\longrightarrow$ ATP + H_2O $\Delta G^{o\prime}$ = 30.5 kJ/mol

Here, the equilibrium constant is the same as the mass-action ratio:

$$K'_{eq} = \frac{[ATP]}{[P_i][ADP]}$$
$$= 5.33 \times 10^2\ M^{-1}.$$

Since $\Delta G = \Delta G^{\circ\prime} + RT \ln K'_{eq}$
$\qquad = 30.5 \text{ kJ/mol} + (2.479 \text{ kJ/mol}) \ln 5.33 \times 10^2$
$\qquad = 46 \text{ kJ/mol}$

14. *Daily ATP Utilization by Human Adults*

(a) A total of 30.5 kJ/mol of free energy is needed to synthesize ATP from ADP and P_i when the reactants and products are at 1 M concentration (standard state). Because the actual physiological concentrations of ATP, ADP, and P_i are not 1 M, the free energy required to synthesize ATP under physiological conditions is different from $\Delta G^{\circ\prime}$. Calculate the free energy required to synthesize ATP in the human hepatocyte when the physiological concentrations of ATP, ADP, and P_i are 3.5, 1.50, and 5.0 mM, respectively.

Answer

$\text{ADP} + P_i \longrightarrow \text{ATP} + H_2O \qquad \Delta G^{\circ\prime} = 30.5 \text{ kJ/mol}$

$K'_{eq} = \dfrac{[\text{ATP}]}{[P_i][\text{ADP}]}$

$\qquad = \dfrac{[3.5 \times 10^{-3} \text{ M}]}{[1.5 \times 10^{-3} \text{ M}][5.0 \times 10^{-3} \text{ M}]}$

$\qquad = 466 \text{ M}^{-1}$

$\Delta G = \Delta G^{\circ\prime} + RT \ln K'_{eq}$
$\qquad = 30.5 \text{ kJ/mol} + (2.479 \text{ kJ/mol}) \ln 466$
$\qquad = 46 \text{ kJ/mol}$

(b) A normal 68 kg (150 lb) adult requires a caloric intake of 2,000 kcal (8,360 kJ) of food per day (24 h). This food is metabolized and the free energy used to synthesize ATP, which is then utilized to do the body's daily chemical and mechanical work. Assuming that the efficiency of converting food energy to ATP is 50%, calculate the weight of ATP utilized by a human adult in a 24 h period. What percentage of body weight does this represent?

Answer

Energy going into ATP synthesis = 8,360 kJ x 50% = 4,180 kJ
Using the value of ΔG from (a), the amount of ATP synthesized is
(4,180 kJ)/(46 kJ/mol) = 91 mol

The molecular weight of ATP = 503 (calculated by summing atomic weights). Thus the weight of ATP synthesized is
(91 mol ATP)(503 g/mol) = 46 kg ATP

As a percentage of body weight:
100(46 kg ATP)/(68 kg body weight) = 68%

(c) Although adults synthesize large amounts of ATP daily, their body weight, structure, and composition do not change significantly during this period. Explain this apparent contradiction.

Answer The concentration of ATP in a healthy body is maintained in a steady state; this is an example of homeostasis, a condition in which the body synthesizes and breaks down ATP as needed.

15. *ATP Reserve in Muscle Tissue*   The ATP concentration in muscle tissue (approximately 70% water) is about 8.0 mM. During strenuous activity each gram of muscle tissue uses ATP at the rate of 300 μmol/min for contraction.

(a) How long would the reserve of ATP last during a 100 meter dash?

(b) The phosphocreatine level in muscle is about 40.0 mM. How does this help extend the reserve of muscle ATP?

(c) Given the size of the reserve ATP pool, how can a person run a marathon?

Answer

(a)   [ATP] = 8.0 x 10^{-3} mol/L = 8.0 x 10^{-6} mol/mL. Because muscle tissue is about 70% water, each gram of muscle contains 0.7 mL of H_2O. Thus the amount of ATP per gram of muscle is
(0.7 mL)(8.0 x 10^{-6} mol/mL) = 5.6 x 10^{-6} mol
Each gram of muscle tissue uses 300 x 10^{-6} mol of ATP per minute during the 100 m dash, thus 5.6 x 10^{-6} mol of ATP would last
$$\frac{(5.6 \times 10^{-6} \text{ mol})(60 \text{ s/min})}{300 \times 10^{-6} \text{ mol/min}} = 1.1 \text{ s}$$

(b)   More ATP can be generated, taking advantage of the high group transfer potential of phosphocreatine:

phosphocreatine + ADP $\longrightarrow$ creatine + ATP $\Delta G^{\circ\prime}$ = -12.5 kJ/mol

(c)   To maintain homeostasis, more ATP can be synthesized, coupled to the catabolism of glycogen, glucose, amino acids, and fatty acids.

16. *Rates of Turnover of γ- and β-Phosphates of ATP*   If a small amount of ATP labeled with radioactive phosphorus in the terminal position, [γ-^{32}P]ATP, is added to a yeast extract, about half of the ^{32}P activity is found in P_i within a few minutes, but the concentration of ATP remains unchanged. Explain. If the same experiment is carried out using ATP labeled with ^{32}P in the central position, [β-^{32}P]ATP, the ^{32}P does not appear in P_i within the same number of minutes. Why?

Answer   To solve this problem, it is useful to represent ATP as A-P-P-P. A radiolabeled phosphate group can be denoted *P. One possible reaction for γ-labeled ATP would be

A-P-P-*P + Glc $\longrightarrow$ A-P-P + G6*P $\longrightarrow \longrightarrow \longrightarrow$ *P_i

or more generally

A-P-P-*P + H_2O $\longrightarrow$ A-P-P + *P_i

These reactions occur relatively quickly. As we saw in Problem 11(c), the concentration of ATP in the body remains in a steady state.

The reaction below occurs more slowly:

A-P-*P-P + H_2O $\longrightarrow$ A-P + *P_i + P_i

As a result, one would expect a faster turnover of *P_i when it is in the γ position than when it is in the β position.

17. *Cleavage of ATP to AMP and PP$_i$ during Metabolism* The synthesis of the activated form of acetate (acetyl-CoA) is carried out in an ATP-dependent process:

$$Acetate + CoA + ATP \longrightarrow acetyl\text{-}CoA + AMP + PP_i$$

(a) The $\Delta G^{\circ\prime}$ for the hydrolysis of acetyl-CoA to acetate and CoA is -32.2 kJ/mol and that for hydrolysis of ATP to AMP and PP$_i$ is -30.5 kJ/mol. Calculate $\Delta G^{\circ\prime}$ for the ATP-dependent synthesis of acetyl-CoA.

(b) Almost all cells contain the enzyme inorganic pyrophosphatase, which catalyzes the hydrolysis of PP$_i$ to P$_i$. What effect does the presence of this enzyme have on the synthesis of acetyl-CoA? Explain.

Answer

(a) The $\Delta G^{\circ\prime}$ can be determined for the coupled reactions:

(1) Acetate + CoA $\longrightarrow$ AcetylCoA + H$_2$O	$\Delta G_1^{\circ\prime}$	= 32.2 kJ/mol
(2) ATP $\longrightarrow$ AMP + PP$_i$	$\Delta G_2^{\circ\prime}$	= -30.5 kJ/mol

Sum: Acetate + ATP $\longrightarrow$ AcetylCoA + AMP + PP$_i$ $\Delta G_{sum}^{\circ\prime}$ = 1.7 kJ/mol

(b) Hydrolysis of PP$_i$ would drive the reaction forward, favoring the synthesis of acetyl-CoA.

18. *Are All Metabolic Reactions at Equilibrium?*

(a) Phosphoenolpyruvate is one of the two phosphate donors in the synthesis of ATP during glycolysis. In human erythrocytes, the steady-state concentration of ATP is 2.24 mM, that of ADP is 0.25 mM, and that of pyruvate is 0.051 mM. Calculate the concentration of phosphoenolpyruvate at 25 °C, assuming that the pyruvate kinase reaction (Fig. 13-3) is at equilibrium in the cell.

(b) The physiological concentration of phosphoenolpyruvate in human erythrocytes is 0.023 mM. Compare this with the value obtained in (a). What is the significance of this difference? Explain.

Answer

(a) First we must calculate the overall $\Delta G^{\circ\prime}$ for the pyruvate kinase reaction, breaking this process into two opposing hydrolyses:

(1) PEP + H$_2$O $\longrightarrow$ pyruvate + P$_i$	$\Delta G_1^{\circ\prime}$	= -61.9 kJ/mol
(2) ADP + P$_i$ $\longrightarrow$ ATP + H$_2$O	$\Delta G_2^{\circ\prime}$	= 30.5 kJ/mol

Sum: PEP + ADP $\longrightarrow$ pyruvate + ATP $\Delta G_{sum}^{\circ\prime}$ = -31.4 kJ/mol

$$
\begin{aligned}
\ln K'_{eq} &= -\Delta G^{\circ\prime}/RT \\
&= -(-31.4 \text{ kJ/mol})/(2.479 \text{ kJ/mol}) \\
&= 12.67 \\
K'_{eq} &= 3.17 \times 10^5
\end{aligned}
$$

Since $K'_{eq} = \dfrac{[pyruvate][ATP]}{[ADP][PEP]}$

$$
\begin{aligned}
[PEP] &= \frac{[pyruvate][ATP]}{[ADP] \, K'_{eq}} \\
&= \frac{(5.1 \times 10^{-5} \text{ M})(2.24 \times 10^{-3} \text{ M})}{(2.5 \times 10^{-4} \text{ M})(3.17 \times 10^5 \text{ M})} \\
&= 1.44 \times 10^{-9} \text{ M}
\end{aligned}
$$

(b)   The physiological [PEP] of 0.023 mM is

$$\frac{0.023 \times 10^{-3} \text{ M}}{1.44 \times 10^{-9} \text{ M}} = 16{,}000 \text{ times the equilibrium concentration}$$

This reaction, like many others in the cell, does not reach equilibrium.

19. *Standard Reduction Potentials*   The standard reduction potential, E'_0, of any redox pair is defined for the half-cell reaction:

Oxidizing agent + n electrons $\rightleftharpoons$ reducing agent

The E'_0 values for the $NAD^+/NADH$ and pyruvate/lactate conjugate redox pairs are -0.32 and -0.19 V, respectively.

(a)   Which conjugate pair has the greater tendency to lose electrons?   Explain.

Answer   The $NAD^+/NADH$ pair is more likely to lose electrons.   The equations in Table 13-7 are written in the direction of reduction (gain of electrons). E'_0 is positive if the oxidized member of a conjugate pair has a tendency to accept electrons. E'_0 is negative if the oxidized member of a conjugate pair does NOT have a tendency to accept electrons. Both $NAD^+/NADH$ and pyruvate/lactate have negative E'_0 values.   The E'_0 of $NAD^+/NADH$ is more negative than that for pyruvate/lactate, so this pair has the greater tendency to accept electrons and is thus the stronger oxidizing system.

(b)   Which is the stronger oxidizing agent?   Explain.

Answer   The pyruvate/lactate pair is the more likely way to accept electrons and thus is the stronger oxidizing agent.   For the same reason that NADH tends to donate electrons to pyruvate, pyruvate tends to accept electrons from NADH.   Pyruvate will be reduced to form lactate; NADH will be oxidized to form NAD^+.   Pyruvate is the oxidizing agent; NADH is the reducing agent.

(c)   If we begin with 1 M concentrations of reactant and product at pH 7, in which direction will the following reaction proceed?

Pyruvate + NADH + H^+ $\rightleftharpoons$ lactate + NAD^+

Answer   From the answers to (a) and (b), it is evident that the reaction will tend to go in the direction of lactate formation.

(d)   What is the standard free-energy change ($\Delta G^{\circ\prime}$) at 25 °C for this reaction?

Answer   The first step is to calculate $\Delta E'_0$ for the reaction, using the E'_0 values from Table 13-7   (Notice that the sign for the $NAD^+/NADH$ pair changes.)

$NADH + H^+ \longrightarrow NAD^+ + 2H^+ + 2e^-$	E'_0	$= 0.320$ V
$\text{Pyruvate} + 2H^+ + 2e^- \longrightarrow \text{lactate}$	E'_0	$= -0.185$ V

$NADH + \text{pyruvate} \longrightarrow NAD^+ + \text{lactate}$	$\Delta E'_0$	$= 0.135$ V

$$\begin{aligned}
\Delta G^{\circ\prime} &= -nF\Delta E'_0 \\
&= -2(96.5 \text{ kJ/mol·V})(0.135 \text{ V}) \\
&= -26 \text{ kJ/mol}
\end{aligned}$$

(e) What is the equilibrium constant (K'_{eq}) for this reaction?

Answer
$$\ln K'_{eq} = -\Delta G^{\circ\prime}/RT$$
$$= -(-26 \text{ kJ/mol})/(2.479 \text{ kJ/mol})$$
$$= -10.49$$
$$K'_{eq} = e^{10.49} = 3.59 \times 10^4$$

20. *Energy Span of the Respiratory Chain* Electron transfer in the mitochondrial respiratory chain may be represented by the net reaction equation

$$\text{NADH} + \text{H}^+ + \frac{1}{2}\text{O}_2 \rightleftharpoons \text{H}_2\text{O} + \text{NAD}^+$$

(a) Calculate the value of $\Delta E'_0$ for the net reaction of mitochondrial electron transfer.

(b) Calculate $\Delta G^{\circ\prime}$ for this reaction.

(c) How many ATP molecules can *theoretically* be generated by this reaction if the standard free energy of ATP synthesis is 30.5 kJ/mol?

Answer
(a) Using E'_0 values from Table 13-7:

$\text{NADH} + \text{H}^+ \longrightarrow \text{NAD}^+ + 2\text{H}^+ + 2e^-$	$E'_0 = 0.320 \text{ V}$
$\frac{1}{2}\text{O}_2 + 2\text{H}^+ + 2e^- \longrightarrow \text{H}_2\text{O}$	$E'_0 = 0.816 \text{ V}$

$\text{NADH} + \text{H}^+ + \frac{1}{2}\text{O}_2 \longrightarrow \text{H}_2\text{O} + \text{NAD}^+$	$\Delta E'_0 = 1.136 \text{ V}$
	$\approx 1.14 \text{ V}$

(b)
$$\Delta G^{\circ\prime} = -nF\Delta E'_0$$
$$= -2(96.5 \text{ kJ/mol·V})(1.14 \text{ V})$$
$$= -220 \text{ kJ/mol}$$

(c) For ATP synthesis, the reaction is

$$\text{ADP} + \text{P}_i \longrightarrow \text{ATP} \qquad \Delta G^{\circ\prime} = 30.5 \text{ kJ/mol}$$

Thus the number of ATP molecules that can, in theory, be generated is
$$\frac{220 \text{ kJ/mol}}{30.5 \text{ kJ/mol}} = 7.2 \approx 7$$

21. *Dependence of Electromotive Force on Concentrations* Calculate the electromotive force (in volts) registered by an electrode immersed in a solution containing the following mixtures of NAD^+ and NADH at pH 7.0 and 25 °C, with reference to a half-cell of $E'_0 = 0.00\text{V}$.

(a) 1.0 mM NAD^+ and 10 mM NADH

(b) 1.0 mM NAD^+ and 1.0 mM NADH

(c) 10 mM NAD^+ and 1.0 mM NADH

Answer The relevant equation for calculating E for this system is
$$E = E'_0 + \frac{RT}{nF} \ln \frac{[\text{NAD}^+]}{[\text{NADH}]}$$

At 25 °C, the (RT/nF) term simplifies to (0.026 V/n).

(a)   From Table 13-7, E'_0 for the $NAD^+/NADH$ redox pair is -0.320 V. Since two
       electrons are transferred, $n = 2$. Thus,
       $$E = (-0.320 \text{ V}) + (0.026 \text{ V}/2) \ln (1 \times 10^{-3} \text{ M})/(10 \times 10^{-3} \text{ M})$$
       $$= -0.35 \text{ V}$$

(b)   The conditions specified here are "standard conditions," so we expect that $E' = E'_0$.
       As proof, we know that the value for $\ln 1 = 0$, so under standard conditions the term
       $(RT/nF) \ln 1 = 0$, which makes $E = E'_0 = -0.320 \text{ V}$.

(c)   Here the concentration of NAD^+ (the electron acceptor) is 10 times that of NADH
       (the electron donor). This affects the value of E'.
       $$E = (-0.320 \text{ V}) + (0.026/2 \text{ V}) \ln (10 \times 10^{-3} \text{ M})/(1 \times 10^{-3})$$
       $$= -0.29 \text{ V}$$

22. *Electron Affinity of Compounds*   List the following substances in order of increasing tendency
to accept electrons: **(a)** α-ketoglutarate + CO_2 (yielding isocitrate), **(b)** oxaloacetate,
(c) O_2, **(d)** $NADP^+$.

Answer   To solve this problem, first write the half-reactions as in Table 13-7, and then find
the value for E'_0 for each. Pay attention to the sign!

Half-reaction	E'_0 (V)
(a) α-Ketoglutarate + CO_2 + $2H^+$ + $2e^- \longrightarrow$ isocitrate	-0.38
(b) Oxaloacetate + $2H^+$ + $2e^- \longrightarrow$ malate	-0.166
(c) $\frac{1}{2} O_2$ + $2H^+$ + $2e^- \longrightarrow H_2O$	0.816
(d) $NADP^+$ + $2H^+$ + $2e^- \longrightarrow$ NADPH + H^+	-0.324

The more positive the E'_0, the more likely is the substance to accept electrons; thus we can
list the substances in order of increasing tendency to accept electrons: (a), (d), (b), (c).

23. *Direction of Oxidation-Reduction Reactions*   Which of the following reactions would be
expected to proceed in the direction shown under standard conditions, assuming that the
appropriate enzymes are present to catalyze them?

(a) Malate + $NAD^+ \longrightarrow$ oxaloacetate + NADH + H^+

(b) Acetoacetate + NADH + $H^+ \longrightarrow \beta$-hydroxybutyrate + NAD^+

(c) Pyruvate + NADH + $H^+ \longrightarrow$ lactate + NAD^+

(d) Pyruvate + β-hydroxybutyrate $\longrightarrow$ lactate + acetoacetate

(e) Malate + pyruvate $\longrightarrow$ oxaloacetate + lactate

(f) Acetaldehyde + succinate $\longrightarrow$ ethanol + fumarate

Answer   It is important to note that standard conditions do not exist in the cell. The value
of $\Delta E'_0$, as calculated in this problem, gives an indication of whether a reaction will or will
not occur in a cell without additional energy being added (usually from ATP); but $\Delta E'_0$ does
not tell the entire story. The actual cellular concentrations of the electron donors and
electron acceptors contribute significantly to the value of E'_0 (see Problem 21, for example).
The potential under nonstandard conditions can either add to an already favorable $\Delta E'_0$, or
can be such a large positive number as to "overwhelm" an unfavorable $\Delta E'_0$ and make ΔE
favorable.

To solve this problem, write the two half-reactions with the appropriate E'_0 values. (see Table 13-7). Pay attention to the sign! Then judge whether the reaction is likely to proceed favorably as written.

(a) Not favorable.

Malate $\longrightarrow$ oxaloacetate $+ 2H^+ + 2e^-$	$E'_0 = 0.166$ V
$NAD^+ + 2H^+ + 2e^- \longrightarrow NADH + H^+$	$E'_0 = -0.320$ V

Malate $+ NAD^+ \longrightarrow$ oxaloacetate $+ NADH$	$\Delta E'_0 = -0.154$ V

(b) Not favorable.

Acetoacetate $+ 2H^+ + 2e^- \longrightarrow \beta$-hydroxybutyrate	$E'_0 = -0.346$ V
$NADH + 2H^+ \longrightarrow NAD^+ + 2H^+ + 2e^-$	$E'_0 = 0.320$ V

Acetoacetate $+ NADH \longrightarrow \beta$-hydroxybutyrate $+ NAD^+$	$\Delta E'_0 = -0.026$ V

(c) Favorable.

Pyruvate $+ 2H^+ + 2e^- \longrightarrow$ lactate	$E'_0 = -0.185$ V
$NADH + H^+ \longrightarrow NAD^+ + 2H^+ + 2e^-$	$E'_0 = 0.320$ V

Pyruvate $+ NADH \longrightarrow$ lactate $+ NAD^+$	$\Delta E'_0 = 0.135$ V

(d) Favorable.

Pyruvate $+ 2H^+ + 2e^- \longrightarrow$ lactate	$E'_0 = -0.185$ V
β-hydroxybutyrate $\longrightarrow$ acetoacetate $+ 2H^+ + 2e^-$	$E'_0 = 0.346$ V

Pyruvate $+ \beta$-hydroxybutyrate $\longrightarrow$ acetoacetate $+$ lactate	$\Delta E'_0 = 0.161$ V

(e) Not favorable.

Malate $\longrightarrow$ oxaloacetate $+ 2H^+ + 2e^-$	$E'_0 = 0.166$ V
Pyruvate $+ 2H^+ + 2e^- \longrightarrow$ lactate	$E'_0 = -0.185$ V

Pyruvate $+$ Malate $\longrightarrow$ lactate $+$ oxaloacetate	$\Delta E'_0 = -0.019$ V

(f) Not favorable.

Acetaldehyde $+ 2H^+ + 2e^- \longrightarrow$ ethanol	$E'_0 = -0.197$ V
Succinate $\longrightarrow$ fumarate $+ 2H^+ + 2e^-$	$E'_0 = -0.031$ V

Acetaldehyde $+$ succinate $\longrightarrow$ ethanol $+$ fumarate	$\Delta E'_0 = -0.228$ V

CHAPTER **14** **Glycolysis and the the Catabolism of Hexoses**

1. *Equation for the Preparatory Phase of Glycolysis* Write balanced equations for all of the reactions in the catabolism of D-glucose to two molecules of D-glyceraldehyde-3-phosphate (the preparatory phase of glycolysis). For each equation write the standard free-energy change. Then write the overall or net equation for the preparatory phase of glycolysis, including the net standard free-energy change.

> *Answer* The initial phase of glycolysis requires ATP; it is endergonic. There are five reactions in this phase (see pages 406-408):
>
> 1. Glucose + ATP $\longrightarrow$ glucose-6-phosphate + ADP $\qquad$ $\Delta G^{\circ\prime}$ = -16.7 kJ/mol
> 2. Glucose-6-phosphate $\longrightarrow$ fructose-6-phosphate $\qquad$ $\Delta G^{\circ\prime}$ = 1.7 kJ/mol
> 3. Fructose-6-phosphate + ATP $\longrightarrow$ fructose-1,6-bisphosphate $\qquad$ $\Delta G^{\circ\prime}$ = -14.2 kJ/mol
> 4. Fructose-1,6-bisphosphate $\longrightarrow$
>   dihydroxyacetone phosphate + glyceraldehyde-3-phosphate $\qquad$ $\Delta G^{\circ\prime}$ = 23.8 kJ/mol
> 5. Dihydroxyacetone phosphate $\longrightarrow$ glyceraldehyde-3-phosphate $\qquad$ $\Delta G^{\circ\prime}$ = 7.5 kJ/mol
>
> The net equation for this phase is
>
> Glucose + 2ATP $\longrightarrow$ 2 glyceraldehyde-3-phosphate + 2ADP + 2H$^+$
>
> The overall standard free-energy change can be calculated by summing the individual reactions: $\Delta G^{\circ\prime}$ = 2.1 kJ/mol (endergonic).

2. *The Payoff Phase of Glycolysis in Skeletal Muscle* In working skeletal muscle under anaerobic conditions, glyceraldehyde-3-phosphate is converted into pyruvate (the payoff phase of glycolysis), and the pyruvate is reduced to lactate. Write balanced equations for all of the reactions in this process, with the standard free-energy change for each. Then write the overall or net equation for the payoff phase of glycolysis (with lactate as the end product), including the net standard free-energy change.

Answer The payoff phase of glycolysis produces ATP, making it is exergonic. This phase consists of five reactions, designated 6-10 in the text (see pp. 409-413):

6. Glyceraldehyde-3-phosphate + P_i + NAD^+ $\longrightarrow$

                1,3-bisphosphoglycerate + NADH + H^+ $\Delta G^{\circ\prime}$ = 6.3 kJ/mol

7. 1,3-Bisphosphoglycerate + ADP $\longrightarrow$

                3-phosphoglycerate + ATP $\Delta G^{\circ\prime}$ = -18.5 kJ/mol

8. 3-Phosphoglycerate $\longrightarrow$ 2-phosphoglycerate $\Delta G^{\circ\prime}$ = 4.4 kJ/mol

9. 2-Phosphoglycerate $\longrightarrow$ phosphoenolpyruvate $\Delta G^{\circ\prime}$ = 7.5 kJ/mol

10. Phosphoenolpyruvate + ADP $\longrightarrow$ pyruvate + ATP $\Delta G^{\circ\prime}$ = -31.4 kJ/mol

The pyruvate is then converted to lactate:

Pyruvate + NADH + H^+ $\longrightarrow$ lactate + NAD^+ $\Delta G^{\circ\prime}$ = -25.1 kJ/mol

The net equation is:

Glyceraldehyde-3-phosphate + 2ADP + P_i $\longrightarrow$ lactate + NAD^+ $\Delta G^{\circ\prime}$ = -56.8 kJ/mol

Since the payoff phase utilizes two glyceraldehyde-3-phosphate molecules from each glucose entering glycolysis, the energetic payoff for the net reaction should be doubled, which makes the net $\Delta G^{\circ\prime}$ = -113.6 kJ/mol.

3. ***Pathway of Atoms in Fermentation*** A "pulse-chase" experiment using ^{14}C-labeled carbon sources is carried out on a yeast extract maintained under strictly anaerobic conditions to produce ethanol. The experiment consists of incubating a small amount of ^{14}C-labeled substrate (the pulse) with the yeast extract just long enough for each intermediate in the pathway to become labeled. The label is then "chased" through the pathway by the addition of excess unlabeled glucose. The "chase" effectively prevents any further entry of labeled glucose into the pathway.

(a) If [1-^{14}C] glucose (glucose labeled at C-1 with ^{14}C) is used as a substrate, what is the location of ^{14}C in the product ethanol? Explain.

(b) Where would ^{14}C have to be located in the starting glucose molecule in order to assure that all the ^{14}C activity were liberated as $^{14}CO_2$ during fermentation to ethanol? Explain.

Answer Anaerobiosis requires the regeneration of NAD^+ from NADH in order to allow glycolysis to continue.

(a) Figure 14-4 illustrates the fate of the carbon atoms of glucose. C-1 (or C-6) becomes C-3 of glyceraldehyde-3-phosphate and subsequently pyruvate. When pyruvate is decarboxylated and reduced to ethanol, C-3 of pyruvate becomes the C-2 of ethanol (*CH_3—CH_2—OH).

(b) In order for all of the labeled carbon from glucose to be converted to $^{14}CO_2$ during ethanol fermentation, the original label would have to be on C-3 and/or C-4 of glucose, since these are converted to the carboxyl group of pyruvate.

4. *Equivalence of Triose Phosphates* ^{14}C-Labeled glyceraldehyde-3-phosphate was added to a yeast extract. After a short time, fructose-1,6-bisphosphate labeled with ^{14}C at C-3 and C-4 was isolated. What was the location of the ^{14}C label in the starting glyceraldehyde-3-phosphate? Where did the second ^{14}C label in fructose-1,6-bisphosphate come from? Explain.

> *Answer* Problem 1 outlines the steps in glycolysis involving fructose-1,6-bisphosphate, glyceraldehyde-3-phosphate, and dihydroxyacetone phosphate. Keep in mind here that the aldolase reaction is readily reversible and the triose phosphate isomerase reaction catalyzes extremely rapid interconversion of its substrates. Thus, label in the C-1 position of glyceraldehyde-3-phosphate would equilibrate with C-1 of dihydroxyacetone phosphate ($\Delta G^{\circ\prime} = 7.5$ kJ/mol). Since the aldolase reaction has a $\Delta G^{\circ\prime} = -23.8$ kJ/mol, favorable in the direction of hexose formation, fructose-1,6-bisphosphate would be readily formed and would be labeled in C-3 and C-4 (see Fig. 14-4).

5. *Glycolysis Shortcut*   Suppose you discovered a mutant yeast whose glycolytic pathway was shorter because of the presence of a new enzyme catalyzing the reaction

Glyceraldehyde-3-phosphate + H_2O + NAD$^+$ $\longrightarrow$ 3-phosphoglycerate + NADH + H$^+$

Although this mutant enzyme shortens glycolysis by one step, how would it affect anaerobic ATP production? Aerobic ATP production?

> *Answer* Under anaerobic conditions, the phosphoglycerate kinase and pyruvate kinase reactions are essential. The shortcut in the mutant yeast would bypass the formation of an acylphosphate by glyceraldehyde-3-phosphate dehydrogenase and therefore would not allow the formation of 1,3-bisphosphoglycerate. Without the formation of a substrate for 3-phosphoglycerate kinase no ATP would be formed. Under anaerobic conditions, the net reaction for glycolysis normally produces 2 ATPs per glucose. In the mutant yeast, net production of ATP would be zero and growth could not occur. Under aerobic conditions, however, since the majority of ATP formation occurs via oxidative phosphorylation, the mutation would have no observable effect.

6. *Role of Lactate Dehydrogenase*   During strenuous activity, muscle tissue demands vast quantities of ATP compared with resting tissue. In rabbit leg muscle or turkey flight muscle, this ATP is produced almost exclusively by lactate fermentation. ATP is produced in the payoff phase of glycolysis by two enzymatic reactions, promoted by phosphoglycerate kinase and pyruvate kinase. Suppose skeletal muscle were devoid of lactate dehydrogenase. Could it carry out strenuous physical activity; that is, could it generate ATP at a high rate by glycolysis? Explain. Remember that the lactate dehydrogenase reaction does not involve ATP. A clear understanding of the answer to this question is essential for comprehension of the glycolytic pathway.

> *Answer* The key point here is that NAD$^+$ must be regenerated from NADH in order for the glycolytic pathway to continue to function. Some tissues, such as skeletal muscle, obtain almost all of their ATP through the glycolytic pathway and are capable of short-term exercise only (see Box 14-1). In order to generate ATP at a high rate, the NADH formed during glycolysis must be oxidized. In the absence of significant amounts of O_2 in the tissues, pyruvate and NADH are converted to lactate and NAD$^+$ by lactate dehydrogenase. In the absence of this enzyme, NAD$^+$ could not be regenerated and glycolytic generation of ATP would stop, and as a consequence, muscle activity could not be maintained.

7. ***Free-Energy Change for Triose Phosphate Oxidation***   The oxidation of glyceraldehyde-3-phosphate to 1,3-bisphosphoglycerate, catalyzed by glyceraldehyde-3-phosphate dehydrogenase, proceeds with an unfavorable equilibrium constant (K'_{eq} = 0.08; $\Delta G°'$ = +6.3 kJ/mol). Despite this unfavorable equilibrium, the flow through this point in the pathway proceeds smoothly. How does the cell overcome the unfavorable equilibrium?

> ***Answer*** In living organisms, where directional flow in a pathway is required, exergonic reactions are coupled to endergonic reactions to overcome unfavorable free-energy changes. The endergonic glyceraldehyde-3-phosphate dehydrogenase reaction is followed by the phosphoglycerate kinase reaction, which rapidly removes the product of the former reaction. Consequently, the dehydrogenase reaction does not reach equilibrium and its unfavorable free-energy change is thus circumvented. The net $\Delta G°'$ of the two reactions, when coupled, is (-18.5 + 6.3) kJ/mol = -12.2 kJ/mol.

8. ***Arsenate Poisoning***   Arsenate is structurally and chemically similar to phosphate (P_i), and many enzymes that require phosphate will also use arsenate. Organic compounds of arsenate are less stable than analogous phosphate compounds, however. For example, acyl arsenates decompose rapidly by hydrolysis in the absence of catalysts:

$$
\begin{array}{c}
\quad\ \ O \quad\quad O \qquad\qquad\qquad\qquad O \qquad\qquad\qquad O \\
\quad\ \ \| \quad\quad\ \| \qquad\qquad\qquad\qquad \| \qquad\qquad\qquad \| \\
R\!-\!C\!-\!O\!-\!As\!-\!O^- \ + \ H_2O \ \longrightarrow \ R\!-\!C\!-\!O^- \ + \ HO\!-\!As\!-\!O^- \ + \ H^+ \\
\qquad\qquad\ \ | \qquad\qquad\qquad\qquad\qquad\qquad\qquad\qquad\ | \\
\qquad\qquad\ \ O^- \qquad\qquad\qquad\qquad\qquad\qquad\qquad\qquad O^-
\end{array}
$$

On the other hand, acyl *phosphates*, such as 1,3-bisphosphoglycerate, are more stable and are transformed in cells by enzymatic action.

(a) Predict the effect on the net reaction catalyzed by glyceraldehyde-3-phosphate dehydrogenase if phosphate were replaced by arsenate.

(b) What would be the consequence to an organism if arsenate were substituted for phosphate? Arsenate is very toxic to most organisms. Explain why.

> ***Answer***
>
> (a)   In the presence of arsenate, the product of the glyceraldehyde-3-phosphate dehydrogenase reaction will be 1-arseno-3-phosphoglycerate, which nonenzymatically decomposes to 3-phosphoglycerate and arsenate, and the substrate for the phosphoglycerate kinase will be bypassed.
>
> (b)   No ATP can be formed in the presence of arsenate because 1,3-bisphosphoglycerate is not formed. Under anaerobic conditions, this would result in no net glycolytic synthesis of ATP. Arsenate poisoning can be used as a test for the presence of an acyl phosphate intermediate in a reaction pathway.

9. **Requirement for Phosphate in Alcohol Fermentation**   In 1906 Harden and Young carried out a series of classic studies on the fermentation of glucose to ethanol and CO_2 by extracts of brewer's yeast and made the following observations. (1) Inorganic phosphate was essential to fermentation; when the supply of phosphate was exhausted, fermentation ceased before all the glucose was used. (2) During fermentation under these conditions, ethanol, CO_2, and a biphosphorylated hexose accumulated. (3) When arsenate was substituted for phosphate, no biphosphorylated hexose accumulated, but the fermentation proceeded until all the glucose was converted into ethanol and CO_2.

(a) Why does fermentation cease when the supply of phosphate is exhausted?

(b) Why do ethanol and CO_2 accumulate? Is the conversion of pyruvate into ethanol and CO_2 essential? Why? Identify the biphosphorylated hexose that accumulates. Why does it accumulate?

(c) Why does the substitution of arsenate for phosphate prevent the accumulation of the biphosphorylated hexose yet allow the fermentation to ethanol and CO_2 to go to completion? (See Problem 8.)

Answer Alcohol fermentation in yeast has the following overall equation

$$\text{Glucose} + 2\text{ADP} + 2\text{P}_i \longrightarrow 2 \text{ ethanol} + 2\text{CO}_2 + 2\text{ATP} + 2\text{H}_2\text{O}$$

It is clear that phosphate is required for the continued operation of glycolysis and ethanol formation. In extracts to which glucose is added, fermentation will proceed until ADP and P_i (present in the extracts) are exhausted.

(a) Phosphate is required in the glyceraldehyde-3-phosphate dehydrogenase reaction, and glycolysis will stop at this step when P_i is exhausted. Since glucose remains it will be phosphorylated by ATP, but P_i will not be released.

(b) Fermentation in yeast cells produces ethanol and CO_2 rather than lactate—see Box 14-3. Without these reactions (in the absence of oxygen), NADH would accumulate and no new NAD^+ would be available for further glycolysis—see Problem 6. The biphosphorylated intermediate (of glycolysis) that accumulates is fructose-1,6-bisphosphate. FBP accumulates because in terms of energetics, it lies at a "low point" or valley in the pathway, between the energy-input reactions that precede it and the energy-payoff reactions that follow.

(c) Arsenate replaces P_i in the glyceraldehyde-3-phosphate dehydrogenase reaction to yield an acylarsenate, which spontaneously hydrolyzes. This prevents the formation of FBP and ATP but allows the formation of 3-phosphoglycerate, which continues through the pathway.

10. **Intracellular Concentration of Free Glucose**   The concentration of glucose in human blood plasma is maintained at about 5 mM. The concentration of free glucose inside muscle cells is much lower. Why is the concentration so low in the cell? What happens to the glucose upon entry into the cell?

Answer Glucose enters cells and is immediately exposed to hexokinase, which converts it to glucose-6-phosphate using the energy of ATP. This reaction is highly exergonic ($\Delta G^{\circ\prime} = -16.7$ kJ/mol), and formation of glucose-6-phosphate is strongly favored. Because the glucose transporter is specific for glucose, glucose-6-phosphate cannot leave the cell and must be stored (as glycogen) or metabolized via glycolysis.

11. *Metabolism of Glycerol* Glycerol (see below) obtained from the breakdown of fat is metabolized by being converted into dihydroxyacetone phosphate, an intermediate in glycolysis, in two enzyme-catalyzed reactions. Propose a reaction sequence for the metabolism of glycerol. On which known enzyme-catalyzed reactions is your proposal based? Write the net equation for the conversion of glycerol to pyruvate based on your scheme.

$$
\begin{array}{c}
\text{OH} \\
| \\
\text{HOCH}_2\text{—C—CH}_2\text{OH} \\
| \\
\text{H}
\end{array}
$$

Answer Glycerol enters metabolism at the center of the glycolytic sequence and is converted to dihydroxyacetone phosphate in two steps, catalyzed by kinase and a dehydrogenase, respectively:

Glycerol + ATP $\longrightarrow$ glycerol-3-phosphate + ADP

Glycerol-3-phosphate + NAD$^+$ $\longrightarrow$ dihydroxyacetone phosphate + NADH + H$^+$

Dihydroxyacetone phosphate is converted to pyruvate in the payoff portion of glycolysis. The overall equation for conversion of glycerol to pyruvate is

Glycerol + 2NAD$^+$ + ADP + P$_i$ $\longrightarrow$ pyruvate + 2NADH + 2H+ + ATP

12. *Measurement of Intracellular Metabolite Concentrations* Measuring the concentrations of metabolic intermediates in the living cell presents a difficult experimental problem. Because cellular enzymes rapidly catalyze metabolic interconversions, a common problem associated with perturbing the cell experimentally is that the measured concentrations of metabolites reflect not the physiological concentrations but the equilibrium concentrations. Hence, a reliable experimental technique requires all enzyme-catalyzed reactions to be instantaneously stopped in the intact tissue, so that the metabolic intermediates do not undergo change. This objective is accomplished by rapidly compressing the tissue between large aluminum plates cooled with liquid nitrogen (-190 °C), a process called **freeze-clamping**. After freezing, which stops enzyme action instantly, the tissue is powdered and the enzymes are inactivated by precipitation with perchloric acid. The precipitate is removed by centrifugation, and the clear supernatant extract is analyzed for metabolites. To calculate the actual intracellular concentration of the metabolite in the cell, the intracellular volume is determined from the total water content of the tissues and a measurement of the extracellular volume.

The actual intracellular concentrations of the substrates and products involved in the phosphorylation of fructose-6-phosphate by the enzyme phosphofructokinase-1 in isolated rat heart tissue are given in the table below.

Metabolite	Apparent concentration (mM)*
Fructose-6-phosphate	0.087
Fructose-1,6-bisphosphate	0.022
ATP	11.42
ADP	1.32

Source: From Williamson, J.R. (1965) Glycolytic control mechanisms I. Inhibition of glycolysis by acetate and pyruvate in the isolated, perfused rat heart. *J. Biol. Chem.* **240**, 2308-2321.

*Calculated as μmol/mL of intracellular water.

(a) Using the information in the table, calculate the mass-action ratio, [fructose-1,6-bisphosphate][ADP]/[fructose-6-phosphate][ATP], for the phosphofructokinase-1 reaction under physiological conditions.

(b) Given that $\Delta G^{\circ\prime}$ for the PFK-1 reaction is -14.2 kJ/mol, calculate the equilibrium constant for this reaction.

(c) Compare the values of the mass-action ratio and K'_{eq}. Is the physiological reaction at equilibrium? Explain. What does this experiment say about the role of PFK-1 as a regulatory enzyme?

Answer

(a) The mass-action ratio is

$$\frac{(0.022 \text{ mM})(1.32 \text{ mM})}{(0.087 \text{ mM})(11.42 \text{ mM})} = 0.029$$

(b) $\Delta G^{\circ\prime}$ = $-RT \ln K'_{eq}$

 $\ln K'_{eq}$ = $-\Delta G^{\circ\prime}/RT$

  = $-(-14.2 \text{ kJ/mol})/(2.479 \text{ kJ/mol})$

  = 5.728

 K'_{eq} = 307

(c) It is clear that the PFK-1 reaction does not approach equilibrium in vivo. This indicates that the product, FBP, does not approximate equilibrium concentrations in vivo, because the pathways are "open systems," operating under near-steady state conditions, with substrates flowing in and products flowing out at all times. Thus, all FBP formed is utilized or turned over rapidly. The PFK-catalyzed reaction, being the rate-limiting step in glycolysis, is thus an excellent candidate for the critical regulatory point of the pathway.

13. *Pasteur Effect* The regulated steps of glycolysis in intact cells are identified by studying the catabolism of glucose in whole tissues or organs. For example, the consumption of glucose by heart muscle can be measured by artificially circulating blood through an isolated intact heart and measuring the concentration of glucose before and after the blood passes through the heart. If the circulating blood is deoxygenated, heart muscle consumes glucose at a steady rate. When oxygen is added to the blood, the rate of glucose consumption drops dramatically, then continues at the new, lower rate. Why?

Answer In the absence of O_2, the ATP needs of the cell are met by anaerobic glucose metabolism (fermentation) to form lactate. This produces a maximum of 2 ATPs per glucose. Because the aerobic metabolism of glucose produces far more ATP per glucose (by oxidative phosphorylation), far less glucose is needed to produce the same amount of ATP. The Pasteur effect was the first demonstration of the primacy of energy production—that is, of ATP levels—in controlling the rate of glycolysis.

14. *Regulation of Phosphofructokinase-1* The effect of ATP on the allosteric enzyme PFK-1 is shown. For a given concentration of fructose-6-phosphate, the PFK-1 activity increases with increasing concentrations of ATP, but a point is reached beyond which increasing concentrations of ATP cause inhibition of the enzyme.

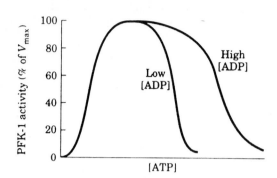

(a) Explain how ATP can be both a substrate and an inhibitor of PFK-1. How is the enzyme regulated by ATP?

(b) In what ways is glycolysis regulated by ATP levels?

(c) The inhibition of PFK-1 by ATP is diminished when the ADP concentration is high [as shown in the graph]. How can this observation be explained?

Answer

(a) In addition to binding sites for substrate(s), allosteric enzymes have binding sites for regulatory metabolites. Binding of effectors to these regulatory sites leads to a modification of enzyme activity by altering its V_{max} or K_m value. ATP is both a substrate and an allosteric inhibitor of PFK-1. Binding of ATP to the catalytic site increases activity, whereas binding to allosteric site inhibits activity.

(b) Because ATP is a negative regulator of PFK-1, elevation of ATP when energy is abundant inhibits the enzyme and thus the flux of metabolites through the glycolytic pathway.

(c) The graph indicates that the addition of ADP suppresses the inhibition of PFK-1 by ATP. Since the total adenine nucleotide pool is fairly constant in all cells, utilization of ATP leads to an increase in ADP. The data indicate that the activity of the enzyme may be regulated in vivo by the ratio of [ATP]/[ADP].

15. *Enzyme Activity and Physiological Function* The V_{max} of the enzyme glycogen phosphorylase from skeletal muscle is much larger than the V_{max} of the same enzyme from liver tissue.

(a) What is the physiological function of glycogen phosphorylase in skeletal muscle? In liver tissue?

(b) Why does the V_{max} of the muscle enzyme need to be larger than that of the liver enzyme?

Answer

(a)   The role of glycogen and its metabolism differs in muscle and liver. *In muscle,* glycogen is broken down to supply energy (ATP), via glycolysis and lactic acid fermentation. Glycogen phosphorylase catalyzes the conversion of stored glycogen to glucose-1-phosphate, which is converted to glucose-6-phosphate and thus enters glycolysis. During strenuous activity, muscle becomes anaerobic and large quantities of glucose-6-phosphate undergo lactic acid fermentation to form the necessary ATP.

In the liver, glycogen is used to maintain the level of glucose in the blood (primarily between meals). In this case, the glucose-6-phosphate is converted to glucose and transported into the bloodstream.

(b)   Strenuous muscular activity requires large amounts of ATP, which must be formed rapidly and efficiently. This requires that glycogen phosphorylase have a high ratio of V_{max}/K_m in muscle. This is not a critical requirement in liver tissue.

16. *Glycogen Phosphorylase Equilibrium*   Glycogen phosphorylase catalyzes the removal of glucose from glycogen. Given that $\Delta G^{\circ\prime}$ for this reaction is 3.1 kJ/mol, calculate the ratio of $[P_i]$ to [glucose-1-phosphate] when this reaction is at equilibrium. (Hint: The removal of glucose units from glycogen does not change the glycogen concentration.) The measured ratio of $[P_i]$ to [glucose-1-phosphate] in muscle cells under physiological conditions is more than 100 to 1. What does this indicate about the direction of metabolite flow through the glycogen phosphorylase reaction? Why are the equilibrium and physiological ratios different? What is the possible significance of this difference?

Answer   First we need to calculate the equilibrium constant from the equation:

$$\Delta G^{\circ\prime} = -RT \ln K'_{eq}$$
$$\ln K'_{eq} = -\Delta G^{\circ\prime}/RT$$
$$= -(3.1 \text{ kJ/mol})/(2.479 \text{ kJ/mol})$$
$$= -1.251$$
$$K'_{eq} = 0.29$$

For the glycogen phosphorylase reaction:

$$\text{Glycogen}_n + P_i = \text{glycogen}_{n-1} + \text{G-1-P}$$

so, $K'_{eq} = \dfrac{(\text{glycogen}_{n-1})(\text{G-1-P})}{(\text{glycogen}_n)(P_i)}$

but since the concentration of glycogen remains constant, these terms cancel and the expression becomes

$$K'_{eq} = \dfrac{\text{G-1-P}}{P_i}$$

This may be arranged to $\dfrac{P_i}{\text{G-1-P}} = \dfrac{1}{K'_{eq}} = \dfrac{1}{0.29} \approx 3.5$

Given that the ration of $[P_i]$ to [glucose-1-P] in muscle cells is greater than 100/1, the rate at which G-1-P can be removed by the mutase is greater than the rate at which it can be produced by glycogen phosphorylase. This indicates that metabolic flow occurs in the direction of glucose to G-1-P, making it likely that the glycogen phosphorylase-catalyzed reaction is the rate-limiting step in glycogen breakdown.

17. **Regulation of Glycogen Phosphorylase**   In muscle tissue, the rate of conversion of glycogen to glucose-6-phosphate is determined by the ratio of phosphorylase *a* (active) to phosphorylase *b* (less active).   Determine what happens to the rate of glycogen breakdown if a muscle preparation containing glycogen phosphorylase is treated with (**a**) phosphorylase *b* kinase and ATP; (**b**) phosphorylase *a* phosphatase; (**c**) epinephrine.

> ***Answer***
> (**a**)   Treatment with the kinase and ATP converts glycogen phosphorylase to the more active, phosphorylated form; glycogen breakdown will accelerate.
>
> (**b**)   Treatment with the phosphatase converts the active phosphorylase *a* to the less active phosphorylase *b*; glycogen breakdown will slow down.
>
> (**c**)   Addition of epinephrine to muscle tissue causes the synthesis of cyclic AMP, which in turn activates the phosphorylase *b* kinase (see Figs. 14-17 and 14-18).   The kinase converts phosphorylase *b* (less active) into phosphorylase *a* (more active); glycogen breakdown will accelerate.

18. **Glycogen Breakdown in Rabbit Muscle**   The intracellular use of glucose and glycogen is tightly regulated at four points.   In order to compare the regulation of glycolysis when oxygen is plentiful and when it is depleted, consider the utilization of glucose and glycogen by rabbit leg muscle in two physiological settings: a resting rabbit, whose leg-muscle ATP demands are low, and a rabbit who has just sighted its mortal enemy, the coyote, and dashes into its burrow at full speed.   For each setting, determine the relative levels (high, intermediate, or low) of AMP, ATP, citrate, and acetyl-CoA and how these levels affect the flow of metabolites through glycolysis by regulating specific enzymes.   In periods of stress, rabbit leg muscle produces much of its ATP by anaerobic glycolysis (lactate fermentation) and very little by oxidation of acetyl-CoA derived from fat breakdown.

> ***Answer*** A primary role of glycolysis is the production of ATP, and the pathway is regulated to ensure efficient ATP formation.   The utilization of glycogen and glucose for energy purposes is regulated at the following steps: glycogen phosphorylase, phosphofructokinase-1, pyruvate kinase, and entry of acetyl CoA into the citric acid cycle.   In muscle, the primary regulatory metabolites are ATP, AMP, citrate, and acetyl CoA.   ATP is an inhibitor of glycogen phosphorylase and PFK-1; AMP stimulates both.   Citrate inhibits PFK-1, and acetyl CoA inhibits pyruvate kinase.   Lack of O_2 leads to elevated levels of NADH, which inhibits pyruvate dehydrogenase and promotes fermentation of pyruvate to lactate.
>
> Under *resting conditions*, ATP levels will be high and AMP levels low since the total adenine nucleotide pools are constant.   Citrate and acetyl CoA levels will be intermediate because O_2 is not limiting and the citric acid cycle will function.   Under conditions of active exertion (running), O_2 becomes limiting and ATP synthesis decreases.   Consequently, [ATP] becomes relatively low and [AMP] becomes relatively high, compared to conditions in the presence of O_2.   Citrate and acetyl CoA are low.   These changes release the inhibition of glycolysis and stimulate lactic acid production.

19. *Glycogen Breakdown in Migrating Birds*   Unlike the rabbit with its short dash, migratory birds require energy for extended periods of time. For example, ducks generally fly several thousand miles during their annual migration. The flight muscles of migratory birds have a high oxidative capacity and obtain the necessary ATP through the oxidation of acetyl-CoA (obtained from fats) via the citric acid cycle. Compare the regulation of muscle glycolysis during short-term intense activity as in the fleeing rabbit, and during extended activity, as in the migrating duck. Why must regulation in these two settings be different?

> *Answer*   Migratory birds have a very efficient respiratory system to ensure that O_2 is available to flight muscles under stress (see Box 14-1). Birds also rely on the aerobic oxidation of fat, since this produces the greatest amount of energy per gram of fuel. Consequently, the migratory bird must regulate glycolysis so that glycogen is used only for short bursts of energy, not for the long-term stress of prolonged flight. The sprinting rabbit relies on breakdown of stored (liver) glycogen plus subsequent anaerobic glycolysis for short-term production of ATP for muscle action.

> The regulation of these two means of ATP production is very different. Under aerobic conditions (see answer to Problem 18) glycolysis is inhibited by ATP levels that remain relatively high, due to feeding acetyl-CoA units derived from fat into the TCA cycle, followed by oxidative phosphorylation. Under anaerobic conditions, glycolysis is stimulated and metabolism of fats does not occur appreciably, since citrate and acetyl-CoA are low and oxygen (the final acceptor of electrons in oxidative phosphorylation) is absent.

20. *Enzyme Defects in Carbohydrate Metabolism*   Summaries of four clinical case studies follow. For each case determine which enzyme is defective and designate the appropriate treatment, from the lists provided. Justify your choices. Answer the questions contained in each case study.

Case A   The patient develops vomiting and diarrhea shortly after milk ingestion. A lactose tolerance test is administered. (The patient ingests a standard amount of lactose, and the blood-plasma glucose and galactose concentrations are measured at intervals. In normal individuals the levels increase to a maximum in about 1 h and then recede.) The patient's blood glucose and galactose concentrations do not rise but remain constant. Explain why the blood glucose and galactose increase and then decrease in normal individuals. Why do they fail to rise in the patient?

Case B   The patient develops vomiting and diarrhea after ingestion of milk. His blood is found to have a low concentration of glucose but a much higher than normal concentration of reducing sugars. The urine gives a positive test for galactose. Why is the reducing-sugar concentration in the blood high? Why does galactose appear in the urine?

Case C   The patient complains of painful muscle cramps when performing strenuous physical exercise but is otherwise normal. A muscle biopsy indicates that muscle glycogen concentration is much higher than in normal individuals. Why does glycogen accumulate?

Case D   The patient is lethargic, her liver is enlarged, and a biopsy of the liver shows large amounts of excess glycogen. She also has a lower than normal level of blood glucose. Account for the low blood glucose concentration in this patient.

> *Defective Enzyme*
> (a) Muscle phosphofructokinase-1
> (b) Phosphomannose isomerase
> (c) Galactose-1-phosphate uridylyltransferase
> (d) Liver glycogen phosphorylase
> (e) Triose kinase
> (f) Lactase in intestinal mucosa
> (g) Maltase in intestinal mucosa

Treatment
1. Jogging 5 km each day
2. Fat-free diet
3. Low-lactose diet
4. Avoiding strenuous exercise
5. Large doses of niacin (the precursor of NAD^+)
6. Frequent and regular feedings

Answer
Case A: (f). Lactase in the intestinal mucosa hydrolyzes milk lactose to glucose and galactose, causing the levels of these sugars to increase transiently after milk ingestion. A patient lacking the enzyme would not exhibit an increase in these sugars but would demonstrate symptoms of lactose toxicity. Such a patient should exclude lactose (milk) from the diet (treatment 3).

Case B: (c). Galactose-1-phosphate uridylyltransferase is an enzyme involved in conversion of galactose to glucose so that the former can enter glycolysis. Absence of this enzyme leads to accumulation of galactose in the blood and excretion in urine. Patients with this deficiency should be on a low-lactose diet (treatment 3).

Case C: (a). This patient must have *lowered* levels of (or a defective) muscle PFK-1 activity, which can function at sufficient rates for all activities except strenuous exercise. This exertion should be avoided (treatment 4).

Case D: (d). Liver glycogen functions as a source of blood glucose. The liver glycogen phosphorylase must be low or defective if glycogen accumulates and blood glucose is low. The patient should eat light meals regularly and frequently (treatment 6).

21. *Severity of Clinical Symptoms Due to Enzyme Deficiency*   The clinical symptoms of the two forms of galactosemia involving the deficiency of galactokinase and galactose-1-phosphate uridylyltransferase show radically different severity. Although both deficiencies produce gastric discomfort upon milk ingestion, the deficiency of the latter enzyme leads to liver, kidney, spleen, and brain dysfunction and eventual death. What products accumulate in the blood and tissues with each enzyme deficiency? Estimate the relative toxicities of these products from the above information.

Answer In galactokinase deficiency, galactose accumulates; in galactose-1-phosphate uridylyltransferase deficiency, galactose-1-phosphate accumulates (see Fig. 14-13). The latter is clearly more toxic.

22. *Preparation of [γ-^{32}P]ATP*   Highly radioactive ATP labeled with ^{32}P in the γ position (terminal phosphate) is used extensively in metabolic studies. In one such procedure, investigators prepared [γ-^{32}P]ATP by incubating the following components:

1 L 50 mM pH 8.0 buffer

10 mmol $MgCl_2$

2 mmol reducing agent (to inhibit disulfide bond formation)

0.4 mmol 3-phosphoglycerate

0.05 mmol NAD^+ (*not* NADH)

0.2 mmol ATP (not radioactive, free of ADP)

0.4 mg glyceraldyde-3-phosphate dehydrogenase

0.2 mg phosphoglycerate kinase

small amount of ^{32}P-labeled sodium phosphate

After the mixture was incubated for 1 h, the ATP was recovered by chromatography. Almost all the ^{32}P was found in the γ position of the ATP. How does this procedure work? Explain the role of all the components except the buffer and the reducing agent.

Answer   The two enzymes, phosphoglycerate kinase (PGK) and glyceraldehyde-3-phosphate dehydrogenase (G3PDH), catalyze the reversible interconversions

3-Phosphoglycerate $\rightleftharpoons$ 1,3-bisphosphoglycerate $\rightleftharpoons$ 3-phosphoglyceraldehyde

These reactions are components of glycolysis. ATP is labeled by the reversible reactions as follows. PGK catalyzes the reversible conversion of

3-phosphoglycerate + ATP $\rightleftharpoons$ 1,3-bisphosphoglycerate + ADP

Although the 1,3-bisphosphoglycerate can bind to G3PDH, but cannot be reduced to glyceraldehyde-3-P because NAD$^+$ (not NADH) is present. However, the phosphate at C-1 (carboxyl) is freely exchanged with the ^{32}P$_i$ in the reaction mixture, due to the formation of the enzyme—S-acyl intermediate (see Fig. 14-5, step 4):

E—SH + R—CO—O—P $\rightleftharpoons$ E—S—CO—R + P$_i$

This "partial" exchange reaction causes formation of 1,3-bisphosphoglycerate with ^{32}P at C-1. When this reacts with the ADP formed in the reversible PGK reaction, γ-labeled ATP is formed.

CHAPTER 15

The Citric Acid Cycle

1. *Balance Sheet for the Citric Acid Cycle* The citric acid cycle uses eight enzymes to catabolize acetyl-CoA: citrate synthase, aconitase, isocitrate dehydrogenase, α-ketoglutarate dehydrogenase, succinyl-CoA synthetase, succinate dehydrogenase, fumarase, and malate dehydrogenase.
   (a) Write a balanced equation for the reaction catalyzed by each enzyme.
   (b) What cofactor(s) are required by each enzyme reaction?
   (c) For each enzyme determine which of the following describes the type of reaction catalyzed: condensation (carbon-carbon bond formation); dehydration (loss of water); hydration (addition of water); decarboxylation (loss of CO_2); oxidation-reduction; substrate-level phosphorylation; isomerization.
   (d) Write a balanced net equation for the catabolism of acetyl-CoA to CO_2.

   Answer
   Citrate synthase
   (a) Acetyl-CoA + oxaloacetate + $H_2O \longrightarrow$ citrate + CoA + H^+
   (b) CoA
   (c) Condensation

   Aconitase:
   (a) Citrate $\longrightarrow$ isocitrate
   (b) No cofactors
   (c) Isomerization

   Isocitrate dehydrogenase
   (a) Isocitrate + $NAD^+ \longrightarrow \alpha$-ketoglutarate + CO_2 + NADH + H^+
   (b) NAD^+
   (c) Oxidative decarboxylation

   α-Ketoglutarate dehydrogenase
   (a) α-ketoglutarate + NAD^+ + CoA $\longrightarrow$ succinyl-CoA + CO_2 + NADH + H^+
   (b) NAD^+, CoA, thiamine pyrophosphate
   (c) Oxidative decarboxylation

   Succinyl CoA synthetase
   (a) Succinyl-CoA + P_i + GDP $\longrightarrow$ succinate + CoA + GTP
   (b) CoA
   (c) Phosphorylation and acyl transfer

Succinate dehydrogenase
(a) Succinate + FAD $\longrightarrow$ fumarate + $FADH_2$
(b) FAD
(c) Oxidation

Fumarase
(a) Fumarate + H_2O $\longrightarrow$ malate
(b) None
(c) Hydration

Malate dehydrogenase
(a) Malate + NAD^+ $\longrightarrow$ oxaloacetate + NADH + H^+
(b) NAD^+
(c) Oxidation
(d) The net equation for the catabolism of acetyl-CoA is

$$Acetyl\text{-}CoA + 3NAD^+ + FAD + GDP + P_i + 2H_2O \longrightarrow$$
$$2CO_2 + CoA + 3NADH + FADH_2 + GTP + 2H^+$$

2. *Recognizing Oxidation and Reduction Reactions in Metabolism* The biochemical strategy of living organisms is the stepwise oxidation of organic compounds to carbon dioxide and water. By properly coupling these reactions, a major part of the energy produced in oxidation is conserved in the form of ATP. It is important to be able to recognize oxidation-reduction processes in metabolism. The reduction of an organic molecule results from the hydrogenation of a double bond (see Eqn 1 below) or of a single bond with accompanying cleavage (Eqn 2). Conversely, the oxidation of an organic molecule results from dehydrogenation. In biochemical redox reactions (see Problem 3) the coenzymes NAD and FAD function to dehydrogenate/hydrogenate organic molecules in the presence of the proper enzymes.

For each of the following metabolic transformations, determine whether oxidation or reduction has occurred. Balance each transformation by inserting H-H, and H_2O where necessary.

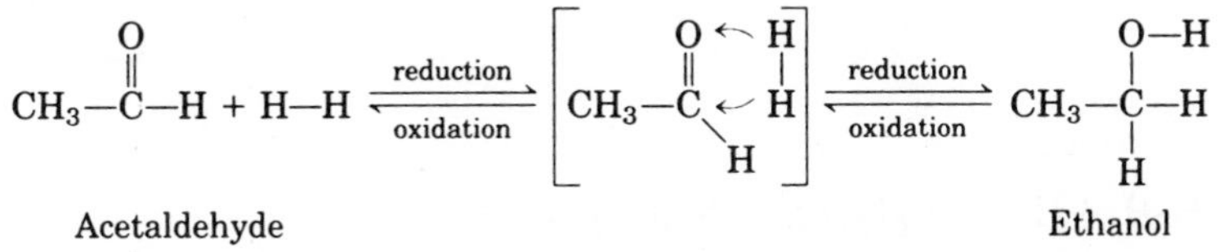

(1)

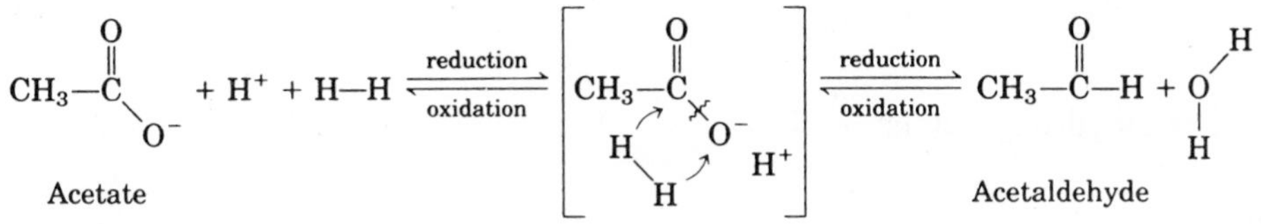

Acetaldehyde   Ethanol

(2)

Acetate   Acetaldehyde

(a) 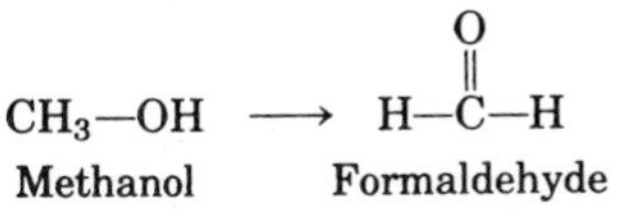 $CH_3-OH \longrightarrow H-\overset{O}{\overset{\|}{C}}-H$
     Methanol   Formaldehyde

(b) 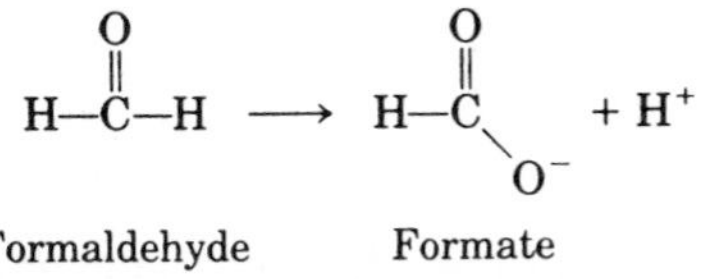 $H-\overset{O}{\overset{\|}{C}}-H \longrightarrow H-\overset{O}{\overset{\|}{C}}\diagdown_{O^-} + H^+$
     Formaldehyde   Formate

(c) 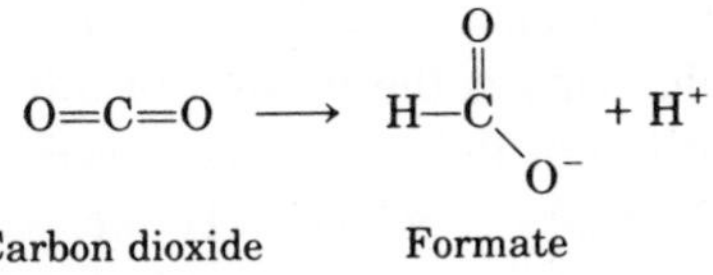 $O=C=O \longrightarrow H-\overset{O}{\overset{\|}{C}}\diagdown_{O^-} + H^+$
     Carbon dioxide   Formate

(d) 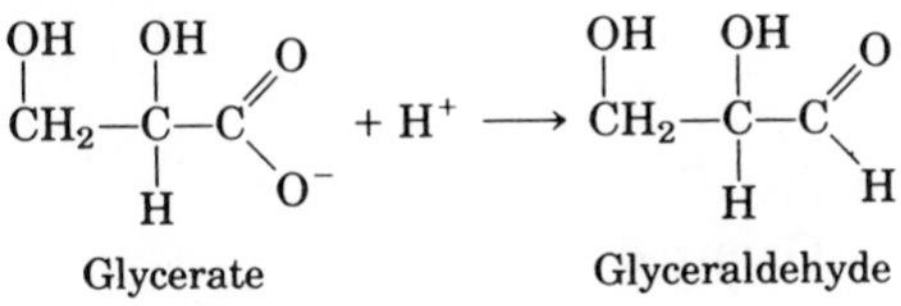 Glycerate $+ H^+ \longrightarrow$ Glyceraldehyde

(e) 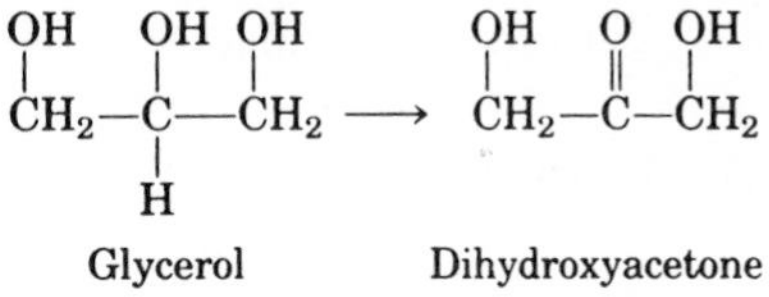 Glycerol $\longrightarrow$ Dihydroxyacetone

(f) 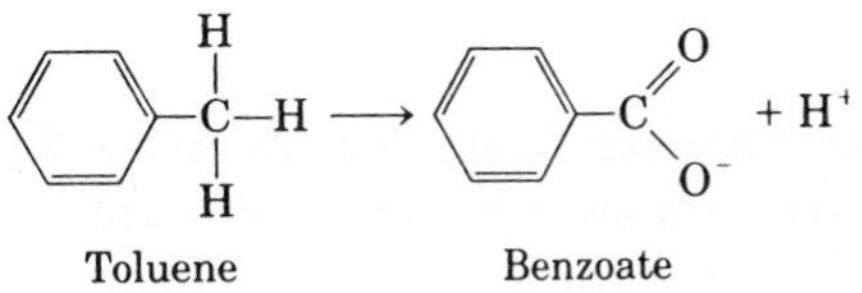 Toluene $\longrightarrow$ Benzoate $+ H^+$

(g) 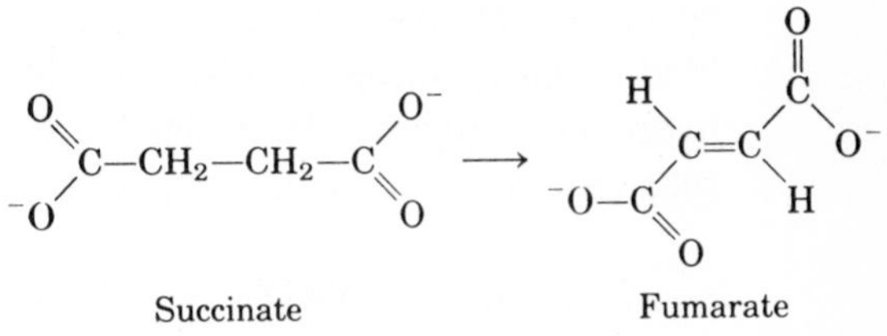 Succinate $\longrightarrow$ Fumarate

(h) Pyruvate $\longrightarrow$ Acetate $+ CO_2$

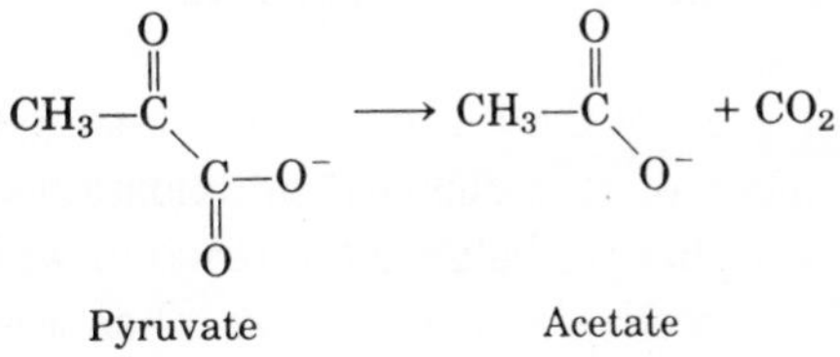

Answer Keep in mind that oxidation is the loss of electrons and accompanying H^+, whereas reduction is the gain of electrons (or H-H).

(a) Oxidation

$$Methanol \longrightarrow formaldehyde + H\text{-}H$$

(b) Oxidation

$$Formaldehyde \longrightarrow formate + H\text{-}H$$

(c) Reduction

$$CO_2 + H\text{-}H \longrightarrow formate + H^+$$

(d) Reduction

$$Glycerate + H\text{-}H + H^+ \longrightarrow glyceraldehyde + H_2O$$

(e) Oxidation

$$Glycerol \rightarrow dihydroxyacetone + H\text{-}H$$

(f) Oxidation

$$Toluene + 2H_2O \rightarrow benzoate + H^+ + 3H\text{-}H$$

(g) Oxidation

$$Succinate \rightarrow fumarate + H\text{-}H$$

(h) Oxidation

$$Pyruvate + H_2O \rightarrow acetate + CO_2 + H\text{-}H$$

3. *Nicotinamide Coenzymes as Reversible Redox Carriers* The nicotinamide coenzymes (see Fig. 13-16) can undergo reversible oxidation-reduction reactions with specific substrates in the presence of the appropriate dehydrogenase. The nicotinamide ring is the portion of the coenzyme involved in the redox reaction; the remaining portion of the coenzyme serves as a binding group recognized by the dehydrogenase protein. Formally, $NADH + H^+$ serves as the hydrogen source (H-H), as described in Problem 2. Whenever the coenzyme is oxidized, a substrate must be simultaneously reduced:

$$Substrate + NADH + H^+ \rightleftharpoons product + NAD^+$$
Oxidized   Reduced   Reduced   Oxidized

For each of the following reactions, determine whether the substrate has been oxidized or reduced or is unchanged in oxidation state (see Problem 2). For substrates that have undergone a redox change, balance the reaction with the necessary amount of NAD^+, NADH, H^+, and H_2O. The objective is to recognize when a redox coenzyme is necessary in a metabolic reaction.

(a) 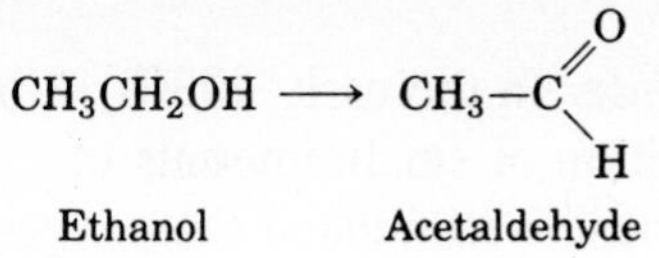 $CH_3CH_2OH \longrightarrow CH_3-C\overset{O}{\underset{H}{\diagdown}}$

Ethanol   Acetaldehyde

(b) 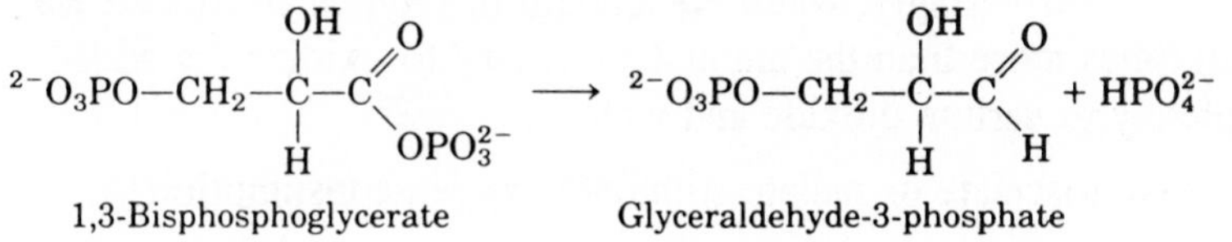 $^{2-}O_3PO-CH_2-\overset{OH}{\underset{H}{C}}-\overset{O}{\underset{OPO_3^{2-}}{C}} \longrightarrow {}^{2-}O_3PO-CH_2-\overset{OH}{\underset{H}{C}}-\overset{O}{\underset{H}{C}} + HPO_4^{2-}$

1,3-Bisphosphoglycerate   Glyceraldehyde-3-phosphate

(c) $CH_3-\overset{O}{\overset{\|}{C}}-\overset{O^-}{\underset{O}{C}} \longrightarrow CH_3-\overset{O^-}{\underset{H}{C}} + CO_2$

Pyruvate   Acetaldehyde

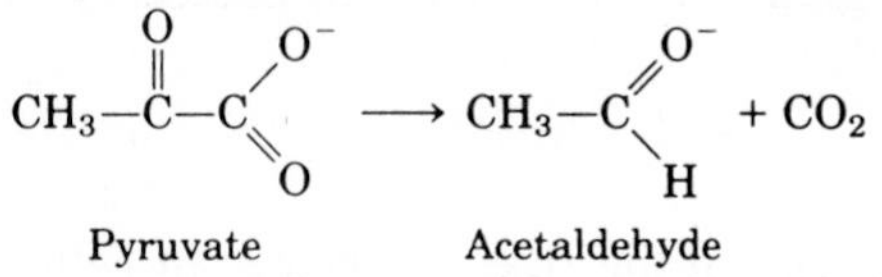

(d) 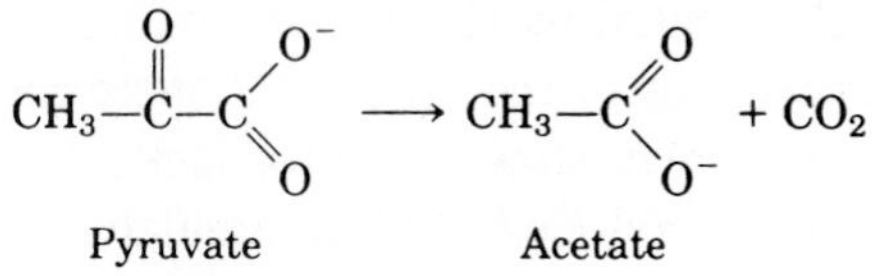 $CH_3-\overset{O}{\overset{\|}{C}}-\overset{O^-}{\underset{O}{C}} \longrightarrow CH_3-\overset{O}{\underset{O^-}{C}} + CO_2$

Pyruvate   Acetate

(e) 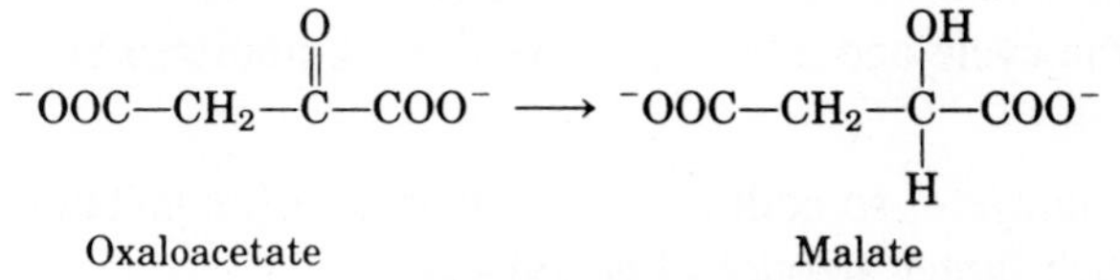 $^-OOC-CH_2-\overset{O}{\overset{\|}{C}}-COO^- \longrightarrow {}^-OOC-CH_2-\overset{OH}{\underset{H}{C}}-COO^-$

Oxaloacetate   Malate

(f) 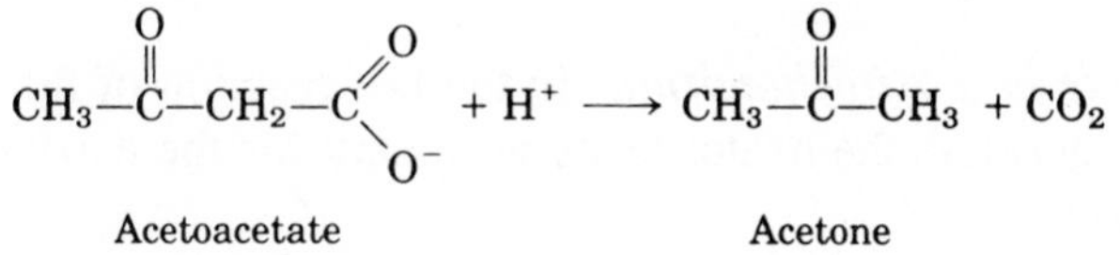 $CH_3-\overset{O}{\overset{\|}{C}}-CH_2-\overset{O}{\underset{O^-}{C}} + H^+ \longrightarrow CH_3-\overset{O}{\overset{\|}{C}}-CH_3 + CO_2$

Acetoacetate   Acetone

Answer

(a) Oxidized

Ethanol + NAD$^+$ $\longrightarrow$ acetaldehyde + NADH + H$^+$

(b) Reduced

1,3-Bisphosphoglycerate + NADH + H$^+$ $\longrightarrow$
glyceraldehyde-3-phosphate + NAD$^+$ + P$_i$

(c) Unchanged

Pyruvate + H$^+$ $\longrightarrow$ acetaldehyde + CO$_2$

(d) Oxidized

Pyruvate + NAD$^+$ $\longrightarrow$ acetate + CO$_2$ + NADH + H$^+$

(e) Reduced

Oxaloacetate + NADH + H$^+$ $\longrightarrow$ malate + NAD$^+$

(f) Unchanged

Acetoacetate + H$^+$ $\longrightarrow$ acetone + CO$_2$

4. ***Stimulation of Oxygen Consumption by Oxaloacetate and Malate*** In the early 1930s, Albert Szent-Györgyi reported the interesting observation that the addition of small amounts of oxaloacetate or malate to suspensions of minced pigeon-breast muscle stimulated the oxygen consumption of the preparation. Surprisingly, when the amount of oxygen consumed was measured, it was about seven times more than the amount necessary to oxidize the added oxaloacetate or malate completely to carbon dioxide and water.

(a) Why does the addition of oxaloacetate or malate stimulate oxygen consumption?

(b) Why is the amount of oxygen consumed so much greater than the amount necessary to oxidize the added oxaloacetate or malate completely?

> *Answer*
> (a) Oxygen consumption is a measure of the activity of the first two stages of cellular respiration: glycolysis and the citric acid cycle. Initial nutrients being oxidized are carbohydrates and lipids. Since several intermediates of the citric acid cycle can be siphoned off into biosynthetic pathways, the cycle may slow down for lack of oxaloacetate in the citrate synthase reaction, and acetyl-CoA will accumulate. Addition of oxaloacetate or malate (converted to oxaloacetate by malate dehydrogenase) will stimulate the cycle and allow it to use the accumulated acetyl-CoA. This stimulates respiration.
>
> (b) Oxaloacetate is regenerated in the cycle, so addition of oxaloacetate (or malate) stimulates the oxidation of a much larger amount of acetyl-CoA.

5. ***The Number of Molecules of Oxaloacetate in a Mitochondrion*** In the last reaction of the citric acid cycle, malate is dehydrogenated to regenerate the oxaloacetate necessary for the entry of acetyl-CoA via the citrate synthase reaction:

$$\text{L-Malate} + NAD^+ \longrightarrow \text{oxaloacetate} + NADH + H^+ \qquad \Delta G^{\circ\prime} = 30 \text{ kJ/mol}$$

(a) Calculate the equilibrium constant for the reaction at 25°C.

(b) Because $\Delta G^{\circ\prime}$ assumes a standard pH of 7, the equilibrium constant obtained in (a) corresponds to

$$K'_{eq} = \frac{[\text{oxaloacetate}][NADH]}{[\text{L-malate}][NADH^+]}$$

The measured concentration of L-malate in rat liver mitochondria is about 0.20 mM when $[NAD^+]/[NADH]$ is 10. Calculate the concentration of oxaloacetate at pH 7 in these mitochondria.

(c) Rat liver mitochondria are roughly spherical, with a diameter of about 2 μm. To appreciate the magnitude of the oxaloacetate concentration in mitochondria, calculate the number of oxaloacetate molecules in a single rat liver mitochondrion.

> *Answer*
> (a) $\Delta G^{\circ\prime} = -RT \ln K'_{eq}$
> $\ln K'_{eq} = -\Delta G^{\circ\prime}/RT$
> $= -(30 \text{ kJ/mol})/(2.479 \text{ kJ/mol})$
> $= -12.10$
> $K'_{eq} = 5.6 \times 10^{-6}$
>
> (b) Since $K'_{eq} = ([OAA][NADH])/([\text{malate}][NAD^+])$
> $[\text{oxaloacetate}] = (K'_{eq}) [\text{malate}][NAD^+]/[NADH]$
> $= (5.6 \times 10^{-6})(2 \times 10^{-4})(10)$
> $= 1.12 \times 10^{-8} \text{ M} \approx 1.1 \times 10^{-8} \text{ M}$

(c)   The volume of a sphere is $4/3 \ \pi r^3$, thus the volume of a mitochondrion is

$$1.33(3.14)(1 \times 10^{-3} \text{ mm})^3 \ = 4.19 \times 10^{-9} \text{ mm}^3$$
$$= 4.19 \times 10^{-15} \text{ L}$$

Given the concentration of oxaloacetate and Avogadro's number, we can calculate the number of molecules in a mitochondrion:

$(1.1 \times 10^{-8} \text{ mol/L})(6.023 \times 10^{23} \text{ molecules/mol})(4.19 \times 10^{-15} \text{ L}) = 28.3 \ \approx \ 28$ molecules

6.   *Respiration Studies in Isolated Mitochondria*   Cellular respiration can be studied using isolated mitochondria and measuring their oxygen consumption under different conditions. If 0.01 M sodium malonate is added to actively respiring mitochondria using pyruvate as a fuel source, respiration soon stops and a metabolic intermediate accumulates.

(a)   What is the structure of the accumulated intermediate?

(b)   Explain why it accumulates.

(c)   Explain why oxygen consumption stops.

(d)   Aside from removing malonate, how can the inhibition of respiration by malonate be overcome? Explain.

Answer   Malonate is a structural analog of succinate and a competitive inhibitor of succinate dehydrogenase.

(a) and (b)   When succinate dehydrogenase is inhibited, succinate accumulates.

(c)   Inhibition of any reaction in a pathway causes the substrate of that reaction to accumulate. Because this substrate is also the product of the preceding reaction, its accumulation changes the effective ΔG of that reaction, and so on for all the steps in the pathway. The net rate of the pathway (or cycle) slows down and eventually becomes almost negligible. In the case of the citric acid cycle, ceasing to produce the primary product, NADH, has the effect of stopping electron transport and consumption of oxygen, the end acceptor of electrons derived from NADH.

(d)   Since malonate is a competitive inhibitor, the addition of large amounts of succinate will overcome the inhibition.

7.   *Labeling Studies in Isolated Mitochondria*   The metabolic pathways of organic compounds have often been delineated by using a radioactively labeled substrate and following the fate of the label.

(a)   How can you determine whether glucose added to a suspension of isolated mitochondria is metabolized to CO_2 and H_2O?

(b)   Suppose you add [3-^{14}C]pyruvate (labeled in the methyl position) to the mitochondria. After one turn of the citric acid cycle, what is the location of the ^{14}C in the oxaloacetate? Explain by tracing the ^{14}C label through the pathway.

(c)   How many turns of the citric acid cycle must the ^{14}C go through before all the isotope is released as $^{14}CO_2$? Explain.

Answer

(a)   The use of uniformly labeled glucose (^{14}C in all carbon atoms) allows one to observe its metabolism. Since glucose is converted to pyruvate, which produces CO_2 in the citric acid cycle, the evolution of $^{14}CO_2$ by mitochondria would confirm the metabolism of glucose to CO_2 and H_2O.

(b)   One turn of the cycle would produce oxaloacetate with label equally distributed between C-2 and C-3 (see Box 15-2).

(c)   Since the second turn of the cycle would release half of the label and every subsequent turn would release another half, it would take an infinite number of turns to release *all* of the labeled carbon.

8.   *1-^{14}C] Glucose Catabolism* If an actively respiring bacterial culture is briefly incubated with [1-^{14}C]glucose and the glycolytic and citric acid cycle intermediates are isolated, where is the ^{14}C in each of the intermediates listed below? Consider only the initial incorporation of ^{14}C into these molecules, in the first pass of labeled glucose through the pathways.

(a) Fructose-1,6-bisphosphate

(b) Glyceraldehyde-3-phosphate

(c) Phosphoenolpyruvate

(d) Acetyl-CoA

(e) Citrate

(f) α-Ketoglutarate

(g) Oxaloacetate

Answer Figures 14-2 and 14-4 and Box 15-2 outline the fate of all the carbon atoms of glucose. If [1-^{14}C] glucose is fed to bacteria for a very short time, the label can be found as follows:

(a) C-1

(b) C-3

(c) C-3

(d) C-2 (methyl group)

(e) Equally distributed in the methylene (—CH_2—) carbons

(f) C-4

(g) Equally distributed in C-2 and C-3, the methylene (—CH_2—) carbons

9.   *Synthesis of Oxaloacetate by the Citric Acid Cycle* Oxaloacetate is formed in the last step of the citric acid cycle by the NAD$^+$-dependent oxidation of L-malate. Can a net synthesis of oxaloacetate take place from acetyl-CoA using only the enzymes and cofactors of the citric acid cycle, without depleting the intermediates of the cycle? Explain. How is the oxaloacetate lost from the cycle (to biosynthetic reactions) replenished?

Answer In the citric acid cycle, the entering acetyl-CoA combines with oxaloacetate to form citrate. One turn of the cycle regenerates oxaloacetate and produces 2 CO_2 molecules. There is *no* net synthesis of oxaloacetate in the cycle. If any cycle intermediates are channeled into biosynthetic reactions, a formation of oxaloacetate is essential. Four enzymes can produce oxaloacetate (or malate) from pyruvate or phosphoenolpyruvate. Pyruvate carboxylase (liver) and PEP carboxykinase (muscle) are the most important in animals, and PEP carboxylase in plants. Malic enzyme produces malate from pyruvate in many organisms (see Table 15-3).

10. *Mode of Action of the Rodenticide Fluoroacetate* Fluoroacetate, prepared commercially for rodent control, is also produced naturally by a South African plant. After entering a cell, fluoroacetate is converted into fluoroacetyl-CoA in a reaction catalyzed by the enzyme acetate thiokinase:

$$F-CH_2COO^- + CoA-SH + ATP \longrightarrow F-CH_2(CO)-S-CoA + AMP + PP_i$$

The toxic effect of fluoroacetate was studied in a metabolic experiment on intact isolated rat heart. After the heart was perfused with 0.22 mM fluoroacetate, the measured rate of glucose uptake and glycolysis decreased and glucose-6-phosphate and fructose-6-phosphate accumulated. An examination of the citric acid cycle intermediates indicated that their concentrations were below normal except for citrate, which had a concentration 10 times higher than normal.

(a) Where does the block in the citric acid cycle occur? What causes citrate to accumulate and the other cycle intermediates to be depleted?

(b) Fluoroacetyl-CoA is enzymatically transformed in the citric acid cycle. What is the structure of the metabolic end product of fluoroacetate? Why does it block the citric acid cycle? How might the inhibition be overcome?

(c) Why do glucose uptake and glycolysis decrease in the heart upon fluoroacetate perfusion? Why do hexose monophosphates accumulate?

(d) Why is fluoroacetate poisoning fatal?

Answer
(a) Fluoroacetate, an analog of acetate, can be activated to fluoroacetyl-CoA and can condense with oxaloacetate to form fluorocitrate. However, it is a strong competitive inhibitor of aconitase.

(b) Fluorocitrate is a structural analog of citrate and a strong competitive inhibitor of aconitase. The inhibition can be overcome by addition of large amounts of citrate.

(c) Both citrate and fluorocitrate are allosteric inhibitors of phosphofructokinase-1. Their accumulation thus inhibits glycolysis and glucose uptake, and causes the accumulation of hexose monophosphates.

(d) The net effect of fluoroacetate is to shut down ATP synthesis, both aerobic (oxidative) and anaerobic (fermentative).

11. *Net Synthesis of α-Ketoglutarate* α-Ketoglutarate plays a central role in the biosynthesis of several amino acids. Write a series of known enzymatic reactions that result in the net synthesis of α-ketoglutarate from pyruvate. Your proposed sequence must not involve the net consumption of other citric acid cycle intermediates. Write the overall reaction for your proposed sequence and identify the source of each reactant.

Answer

Anaplerotic reactions replenish lost intermediates in the citric acid cycle. Synthesis of α-ketoglutarate from pyruvate occurs by the sequential action of pyruvate carboxylase (which makes extra molecules of oxaloacetate); plus pyruvate dehydrogenase and the citric acid cycle enzymes citrate synthase, aconitase, and isocitrate dehydrogenase. These make up a series of reactions.

$$\text{Pyruvate} + \text{ATP} + CO_2 \longrightarrow \text{oxaloacetate} + \text{ADP} + P_i$$
$$\text{Pyruvate} + \text{NAD}^+ + \text{CoA} \longrightarrow \text{Ac-CoA} + CO_2 + \text{NADH} + H^+$$
$$\text{Oxaloacetate} + \text{Ac-CoA} \longrightarrow \text{citrate} + \text{CoA}$$
$$\text{Citrate} \longrightarrow \text{Isocitrate}$$
$$\text{Isocitrate} + \text{NAD}^+ \longrightarrow \alpha\text{-ketoglutarate} + \text{NADH} + H^+ + CO_2$$

Net reaction:

$$2\text{ Pyruvate} + \text{ATP} + 2\text{NAD}^+ + H_2O \longrightarrow$$
$$\alpha\text{-ketoglutarate} + CO_2 + 2\text{NADH} + \text{ADP} + P_i + 3H^+$$

12. *Regulation of Citrate Synthase* In the presence of saturating amounts of oxaloacetate, the activity of citrate synthase from pig heart tissue shows a sigmoid dependence on the concentration of acetyl-CoA, as shown below. When succinyl-CoA is added, the curve shifts to the right and becomes even more sigmoid.

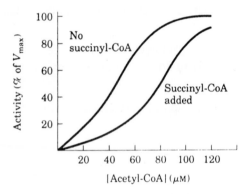

On the basis of these observations, explain how succinyl-CoA regulates the activity of citrate synthase (Hint: See Fig. 8-27). Why is succinyl-CoA an appropriate signal for regulation of the citric acid cycle? How does the regulation of citrate synthase control the rate of cellular respiration in pig heart tissue?

Answer

Succinyl-CoA is the first 4-C intermediate of the citric acid cycle, occurring just prior to steps that form $FADH_2$ and cause overall transformation to oxaloacetate. Accumulation of ATP (a negative feedback signal) results in decreased rates of $FADH_2$ oxidation by the electron transport chain, which would in turn cause accumulation of succinyl-CoA, making this intermediate an appropriate signal from within the citric acid cycle for events occurring "downstream."

The curves in the figure (p. 478) show that succinyl-CoA shifts the half-saturation point ($S_{0.5}$) of the sigmoidal curve for Ac-CoA to the right, but does not alter V_{max}. This indicates that it may act directly as a competitive inhibitor vs. Ac-CoA or by binding to a site separate from the active site.

Citrate synthase catalyzes the "first" step in the citric acid cycle, the entry point for input of Ac-CoA. In the absence of sufficient uptake of Ac-CoA by the cycle, it can instead be shunted toward fatty acid synthesis.

13. *Regulation of Pyruvate Carboxylase* The carboxylation of pyruvate by pyruvate carboxylase occurs at a very low rate unless acetyl-CoA, a positive allosteric modulator, is present. If you have just completed a meal rich in fatty acids (triacylglycerols) but low in carbohydrates (glucose) how does this regulatory property shut down the oxidation of glucose to CO_2 and H_2O but increase the oxidation of acetyl-CoA derived from fatty acids?

   Answer Fatty acid catabolism increases the level of acetyl-CoA, which stimulates pyruvate carboxylase. The resulting increase in oxaloacetate stimulates acetyl-CoA consumption through the citric acid cycle, causing the citrate and ATP concentrations to rise. These metabolites inhibit glycolysis at PFK-1 and inhibit pyruvate dehydrogenase, effectively slowing the utilization of sugars and pyruvate.

14. *Relationship between Respiration and the Citric Acid Cycle* Although oxygen does not participate directly in the citric acid cycle, the cycle operates only when O_2 is present. Why?

   Answer Oxygen is the terminal electron acceptor in oxidative phosphorylation, and thus is needed to recycle NAD^+ from NADH. NADH is produced in greatest quantities by the oxidative reactions of the citric acid cycle. In the absence of O_2, NADH accumulates and allosterically inhibits pyruvate dehydrogenase and α-ketoglutarate dehydrogenase (see Fig. 15-14).

15. *Thermodynamics of Citrate Synthase Reaction In Vivo* Citrate is formed by the citrate synthase-catalyzed condensation of acetyl-CoA with oxaloacetate:

   $$\text{Oxaloacetate + acetyl-CoA} + H_2O \longrightarrow \text{citrate + CoA} + H^+$$

In rat heart mitochondria at pH 7.0 and 25 °C, the concentrations of reactants and products are: oxaloacetate, 1 μM; acetyl-CoA, 1 μM; citrate, 220 μM; and CoA, 65 μM. On the basis of these concentrations and the value of the standard free-energy change for the citrate synthase reaction (-32.2 kJ/mol), determine the direction of metabolite flow through the citrate synthase reaction in the cell. Explain.

   Answer The free-energy change of the citrate synthase reaction in the cell is

   $$\Delta G = \Delta G^{\circ\prime} + RT \ln \frac{[\text{citrate}][\text{CoA}]}{[\text{OAA}][\text{Acetyl-CoA}]}$$

   $$= -32.2 \text{ kJ/mol} + (2.479 \text{ kJ/mol}) \ln \frac{(220 \times 10^{-6} \text{ M})(65 \times 10^{-6} \text{ M})}{(1 \times 10^{-6} \text{ M})(1 \times 10^{-6} \text{ M})}$$

   $$= -8.48 \text{ kJ/mol}$$

Thus the citrate synthase reaction is exergonic and proceeds in the direction of citrate formation.

16. ***Reactions of the Pyruvate Dehydrogenase Complex*** Two of the steps in the oxidative decarboxylation of pyruvate (steps 4 and 5, Fig. 15-6) do not involve any of the three carbons of pyruvate yet are essential to the operation of the pyruvate dehydrogenase complex. Explain.

Answer The pyruvate dehydrogenase complex can be thought of as performing five enzymatic reactions. The first three (see Fig. 15-6) catalyze the oxidation of pyruvate to acetyl-CoA and reduction of the enzyme. The last two reactions convert reduced enzyme and NAD^+ to oxidized enzyme and $NADH + H^+$. The moiety on the enzyme that is oxidized/reduced is the lipoamide cofactor.

CHAPTER **16**   **Oxidation of Fatty Acids**

1. *Energy in Triacylglycerols* On a per-carbon basis, where does the largest amount of biologically available energy in triacylglycerols reside: in the fatty acid portions or the glycerol portion? Indicate how knowledge of the chemical structure of triacylglycerols provides the answer.

    Answer The fatty acids of triacylglycerols are hydrocarbon in nature (except for the carboxyl group) and require a great deal of oxidation for complete catabolism. Glycerol on the other hand is partially oxidized, having an —OH group on each carbon. Thus, the fatty acids produce far more energy per carbon during oxidation than does glycerol. Triacylglycerols have an energy of oxidation more than twice that for the same weight of carbohydrates or proteins.

2. *Fuel Reserves in Adipose Tissue* Triacylglycerols have the highest energy content of the major nutrients.

    (a) If 15% of the body mass of a 70 kg adult consists of triacylglycerols, calculate the total available fuel reserve, in both kilojoules and kilocalories, in the form of triacylglycerols. Recall that 1.00 kcal = 4.18 kJ, and that 1.0 kcal = 1.0 nutrional Calorie.

    (b) If the basal energy requirement is approximately 8,400 kJ/day (2,000 kcal/day) how long could this person survive if the oxidation of fatty acids stored as triacylglycerols were the only source of energy?

    (c) What would be the weight loss per day in pounds under such starvation conditions (1 lb = 0.454 kg)?

    Answer
    (a) Knowing that the energy value of stored triacylglycerol is 38 kJ/g, the available fuel reserve is

    $$(0.15)(70 \times 10^3 \text{ g})(38 \text{ kJ/g}) = 4 \times 10^5 \text{ kJ}$$
    $$= 9.5 \times 10^4 \text{ kcal}$$

    (b) At a rate of 8.4×10^3 kJ/day, the fuel supply would last

    $$(4 \times 10^5 \text{ kJ})/(8.4 \times 10^3 \text{ kJ/day}) = 48 \text{ days}$$

    (c) If the total triacylglycerol is used over a 48 day period, this represents a rate of weight loss of

    $$\frac{(0.15)(70 \times 10^3 \text{ g})}{48 \text{ days}} = 218 \text{ g/day} \approx 0.5 \text{ lb/day}$$

3. ***Common Reaction Steps in the Fatty Acid Oxidation Cycle and Citric Acid Cycle*** Cells often follow the same enzyme reaction pattern for bringing about analogous metabolic reactions. For example, the steps in the oxidation of pyruvate and α-ketoglutarate to acetyl-CoA and succinyl-CoA, although catalyzed by different enzymes, are very similar. The first stage in the oxidation of fatty acids follows a reaction sequence closely resembling one in the citric acid cycle. Show by equations the analogous reaction sequences in the two pathways.

> ***Answer*** The first three reactions in the β-oxidation of fatty acyl-CoA molecules are analogous to three reactions of the citric acid cycle.
>
> The fatty acyl-CoA dehydrogenase reaction is analogous to the succinate dehydrogenase reaction, as both are FAD-requiring oxidations:
>
> > Succinate + FAD $\longrightarrow$ fumarate + FADH$_2$
> > Fatty acyl-CoA + FAD $\longrightarrow$ fatty acyl-trans-Δ^2-enoyl-CoA + FADH$_2$
>
> The enoyl-CoA hydratase reaction is analogous to the fumarase reaction (both add water to an olefinic bond):
>
> > Fumarate + H$_2$O $\longrightarrow$ malate
> > Fatty acyl-trans-Δ^2-enoyl-CoA + H$_2$O $\longrightarrow$ L-β-hydroxy acyl-CoA
>
> The β-hydroxy acyl-CoA dehydrogenase reaction is analogous to the malate dehydrogenase reaction (both are NAD-requiring and act upon β-hydroxy acyl compounds):
>
> > Malate + NAD$^+$ $\longrightarrow$ oxalacetate + NADH
> > L-β-Hydroxy acyl-CoA + NAD$^+$ $\longrightarrow$ β-ketoacyl-CoA + NADH

4. ***The Chemistry of the Acyl-CoA Synthetase Reaction*** Fatty acids are converted into their coenzyme A esters by the reversible reaction catalyzed by acyl-CoA synthetase:

> R—COO$^-$ + ATP + CoA $\rightleftharpoons$ R—(CO)—CoA + AMP + PP$_i$

(a) The enzyme-bound intermediate in this reaction has been identified as the mixed anhydride of the fatty acid and adenosine monophosphate (AMP), acyl-AMP:

Write two equations corresponding to the two steps involved in the reaction catalyzed by acyl-CoA synthetase.

(b) The reaction above is readily reversible, with an equilibrium constant near 1. How can this reaction be made to favor formation of fatty acyl-CoA?

> ***Answer*** The activation of carboxyl groups by ATP could in theory be accomplished by three types of reactions: the formation of acyl-phosphate + ADP, or acyl-ADP + P$_i$, or acyl-AMP + PP$_i$.

All of these reactions are readily reversible. In order to create an activation reaction with a highly negative $\Delta G^{\circ\prime}$ (effectively irreversible), the third type of reaction (only) can be coupled to a pyrophosphatase, as in the synthesis of fatty acyl-CoA molecules.

(a) $R—COOH + ATP \longrightarrow$ acyl-AMP + PP_i
      Acyl-AMP + CoA $\longrightarrow$ acyl-CoA + AMP

(b)   The hydrolysis of PP_i by an inorganic pyrophosphatase

5.   *Oxidation of Tritiated Palmitate* Palmitate uniformly labeled with tritium (^3H) to a specific activity of 2.48 x 10^8 counts per minute (cpm) per micromole of palmitate is added to a mitochondrial preparation that oxidizes it to acetyl-CoA. The acetyl-CoA is isolated and hydrolyzed to acetate. The specific activity of the isolated acetate is 1.00 x 10^7 cpm per micromole. Is this result consistent with the β-oxidation pathway? Explain. What is the final fate of the removed tritium?

   Answer The β-oxidation pathway includes two dehydrogenase enzymes that remove hydrogen (H-H) from the fatty acyl-CoA chain at both $-CH_2-CH_2-$ and CH-OH structures (see answer to Problem 2). The net result of these two reactions is removal of one of the two hydrogens present at the point of formation of the enoyl-CoA intermediate. The two other H's that appear in the methyl group of acetyl-CoA come from water. If palmitate contains 16 carbons, each containing (14 x 2) + 3 = 31 H's, so each 2-carbon unit contains roughly 4 or 1/8 of the total ^3H present. Thus, the amount expected to be present per acetyl-CoA is 0.5(2.48 x 10^8 cpm)/8 = 1.55 x 10^7 cpm, somewhat higher than observed. Exchange with solvent water from B-ketoacyl-CoA could cause additional loss of ^3H.

6.   *Compartmentation in β Oxidation* Free palmitate is activated to its coenzyme A derivative (palmitoyl-CoA) in the cytosol before it can be oxidized in the mitochondrion. If palmitate and [^{14}C]coenzyme A are added to a liver homogenate, palmitoyl-CoA isolated from the cytosolic fraction is radioactive, but that isolated from the mitochondrial fraction is not. Explain.

   Answer The transport of fatty acid molecules into mitochondria requires a shuttle system involving a fatty acyl-carnitine intermediate. Fatty acids are first converted to fatty acyl-CoA molecules by the action of a kinase in the cytosol, followed by (at the inner membrane of the mitochondrion), transfer to carnitine and release of CoA. After transport of fatty acyl-carnitine through the membrane, the fatty acyl group is transferred to mitochondrial CoA and is oxidized. The cytosolic and mitochondrial pools of CoA are kept separate, and no labeled CoA from the cytosolic pool will enter the mitochondrion.

7.   *Effect of Carnitine Deficiency* A patient developed a condition characterized by progressive muscular weakness and aching muscle cramps. These symptoms were aggravated by fasting, exercise, and a high-fat diet. The homogenate of a muscle specimen from the patient oxidized added oleate more slowly than did control homogenates of muscle specimens from healthy individuals. When carnitine was added to the patient's muscle homogenate, the rate of oleate oxidation equaled that in the control homogenates. The patient was diagnosed as having a carnitine deficiency.

(a) Why did added carnitine increase the rate of oleate oxidation in the patient's muscle homogenate?

(b) Why were the symptoms aggravated by fasting, exercise, and a high-fat diet?

(c) Suggest two possible reasons for the deficiency of muscle carnitine in the patient.

> **Answer**
>
> (a) The carnitine-mediated transport of fatty acids into mitochondria is the rate-limiting step in β oxidation (see Fig. 16-6). Carnitine deficiency decreases the rate of transport of fatty acids into mitochondria and thus the rate of β oxidation, so addition of carnitine would increase the rate of oxidation.
>
> (b) Fasting, exercise, and a high fat diet cause an increased need for β oxidation of fatty acids. These conditions increase the need for the carnitine shuttle system and thus the severity of the symptoms of carnitine deficiency.
>
> (c) The deficiency of carnitine may result from a deficiency of its precursor (lysine) or from a defect in one of the enzymes that synthesizes carnitine from this precursor.

8. ***Fatty Acids as a Source of Water*** Contrary to legend, camels do not store water in their humps, which actually consist of a large fat deposit. How can these fat deposits serve as a source of water? Calculate the amount of water (in liters) that can be produced by the camel from 1 kg (0.45 lb) of fat. Assume for simplicity that the fat consists entirely of tripalmitoylglycerol.

> **Answer** In addition to energy and CO_2, oxidation of fatty acids produces water in significant amounts. Oxidation of palmityl-CoA involves:
>
> $$CH_3\text{-}(CH_2)_{14}\text{-}CO\text{-}SCoA + 23O_2 + 131P_i + 131ADP \longrightarrow$$
> $$CoA\text{-}SH + 131ATP + 16CO_2 + 146H_2O$$
>
> Subtracting the H_2O formed as part of ATP synthesis (131 mol), oxidizing one mol of palmitate yields an additional 15 mol of water. Since one mol of tripalmitylglycerol contains 3 mol of palmitate, oxidizing 1 mol of triacylglycerol produces 45 mol of water. Using a value of $M_r = 802$ g/mol for fat, it can be calculated that 1 kg of fat produces
>
> $$\frac{(1000 \text{ g})(45 \text{ mol})(18 \text{ g/mol})}{(802 \text{ g})} \approx 1000 \text{ g} = 1 \text{ L of water}$$
>
> Note: This may be an overestimate, since the fatty acyl groups on the triacylglycerol may be less highly reduced than palmitate and because of the weight contribution of the glycerol component.

9. ***Petroleum as a Microbial Food Source*** Some microorganisms of the genera *Nocardia* and *Pseudomonas* can grow in an environment where hydrocarbons are the only food source. These bacteria oxidize straight-chain aliphatic hydrocarbons, for example, octane, to their corresponding carboxylic acids:

$$CH_3(CH_2)_6CH_3 + NAD^+ + O_2 \longrightarrow CH_3(CH_2)_6COOH + NADH + H^+$$

How can these bacteria be used to clean up oil spills?

> **Answer** The oxidation of hydrocarbons to the corresponding fatty acids allows these microbes to obtain all of the energy needed for growth from β oxidation and oxidative phosphorylation. Thus the hydrocarbons can be converted to CO_2 and H_2O. Theoretically, the oil in spills could be broken down by treatment with these microbes.

10. **Metabolism of a Straight-Chain Phenylated Fatty Acid** A crystalline metabolite was isolated from the urine of a rabbit that had been fed a straight-chain fatty acid containing a terminal phenyl group:

$$\text{\Large\bigcirc}\!\!-CH_2-(CH_2)_n-COO^-$$

A 302 mg sample of the metabolite in aqueous solution was completely neutralized by adding 22.2 mL of 0.1 M NaOH.

(a) What is the probable molecular weight and structure of the metabolite?

(b) Did the straight-chain fatty acid fed to the rabbit contain an even or an odd number of methylene ($-CH_2-$) groups (i.e., is n even or odd)? Explain.

Answer

(a)   22.2 mL of 0.1 M NaOH is equivalent to $(22.3 \times 10^{-3} \text{ L})(0.1 \text{ mol/L}) = 22.2 \times 10^{-4}$ mol of unknown metabolite (assuming that it contains only one carboxyl group). Thus the M_r of the metabolite is

$$\frac{302 \times 10^{-3} \text{ g}}{22.2 \times 10^{-4} \text{ mol}} \quad = \quad 136$$

This is the M_r of phenylacetate.

(b)   Since β oxidation removes two-carbon units, and since the end product is a two-carbon unit, the original fatty acyl chain must have been even-numbered (with the phenyl group counted as equivalent to a terminal methyl group). An odd-numbered fatty acid would have produced phenylpropionate.

11. **Fatty Acid Oxidation in Diabetics** When the acetyl-CoA produced during β oxidation in the liver exceeds the capacity of the citric acid cycle, the excess acetyl-CoA reacts to form the ketone bodies acetoacetate, D-β-hydroxybutyrate, and acetone. This condition exists in cases of severe diabetes because the patient's tissues cannot use glucose; they oxidize large amounts of fatty acids instead. Although acetyl-CoA is not toxic, the mitochondrion must divert the acetyl-CoA to ketone bodies. Why? How does this diversion solve the problem?

Answer Diabetics oxidize large quantities of fat because of their inability to use glucose efficiently. This leads to a decrease in the activity of the citric acid cycle (see Problem 12; see also Chapter 15, Problem 9) and an increase in the pool of acetyl-CoA. Because the mitochondrial CoA pool is small, liver mitochondria recycle CoA by condensing two acetyl-CoA molecules to form acetoacetyl-CoA + CoA (see Fig. 16 -16). The acetoacetyl-CoA is converted to other ketones, and the CoA is recycled for use in the β-oxidation pathway and energy production.

12. **Consequences of a High-Fat Diet with No Carbohydrates** Suppose you had to subsist on a diet of whale and seal blubber with little or no carbohydrate.

(a) What would be the effect of carbohydrate deprivation on the utilization of fats for energy?

(b) If your diet were totally devoid of carbohydrate, would it be better to consume odd- or even-numbered fatty acids? Explain.

Answer

(a)   Pyruvate, formed from glucose via glycolysis, is the main source of the oxaloacetate needed to replenish citric acid cycle intermediates (see Table 15-3). In the absence of carbohydrate in the diet, the oxaloacetate level drops and the citric acid cycle slows. This increases the rate of β oxidation of fatty acids and leads to ketosis.

(b)   The last cycle of β oxidation produces 2 acetyl-CoA molecules from an even-numbered fatty acid, or propionyl-CoA + acetyl-CoA from an odd-numbered fatty acid. Propionyl-CoA can be converted to succinyl-CoA (see Fig. 16 -12), which when converted to oxaloacetate stimulates the citric acid cycle and relieves the conditions leading to ketosis. Thus it would be better to consume odd-numbered fatty acids.

13. *Formation of Acetyl-CoA from Fatty Acid Precursors*   Write a balanced net equation for the formation of acetyl-CoA from the following substances, including all activation steps:

   (a) Myristoyl-CoA

   (b) Stearate

   (c) D-β-Hydroxybutyrate

   Answer

   (a) Myristoyl-CoA + 6CoA + 6FAD + 6NAD$^+$ + 6H$_2$O $\longrightarrow$
$$7 \text{ acetyl-CoA} + 6\text{FADH}_2 + 6\text{NADH} + 6\text{H}^+$$

   (b) Stearate + ATP + 9CoA + 8FAD + 8NAD$^+$ + 9H$_2$O $\longrightarrow$
$$9 \text{ acetyl-CoA} + \text{AMP} + \text{PP}_i + 8\text{FADH}_2 + 8\text{NADH} + 10\text{H}^+$$

   (c) β-Hydroxybutyrate + ATP + 2CoA + NAD$^+$ + H$_2$O $\longrightarrow$
$$2 \text{ acetyl-CoA} + \text{AMP} + \text{PP}_i + \text{NADH} + 3\text{H}^+$$

14. *Pathway of Labeled Atoms during Fatty Acid Oxidation*   [9-^{14}C]Palmitate is oxidized under conditions in which the citric acid cycle is operating. What will be the location of ^{14}C in **(a)** acetyl-CoA, **(b)** citrate, and **(c)** butyryl-CoA? Assume only one turn of the citric acid cycle.

   Answer

   (a)   After four cycles of β oxidation, [9-^{14}C]palmitate is converted to 4 acetyl-CoA + [1-^{14}C]octanoate. The next cycle will remove an acetyl-CoA labeled in the carboxyl carbon.

   (b)   [1-^{14}C]Acetate released in the fifth cycle condenses with oxaloacetate to form citrate with the label in one of the two terminal carboxyl groups.

   (c)   Butyryl-CoA is formed after six cycles of β oxidation of palmitate. The C-9 label in palmitate was converted to [1-^{14}C] acetyl-CoA in the fifth cycle, so the butyryl-CoA is not labeled.

15. *Net Equation for Complete Oxidation of D-β-Hydroxybutyrate*   Write the net equation for the complete oxidation of D-β-hydroxybutyrate in the kidney. Include any required activation steps and all oxidative phosphorylations.

   Answer   The net equation is (see Table 15-2, Fig. 16-8, and Eqn 16-5)

   β-Hydroxybutyrate + 4½ O$_2$ + 25ADP + 25P$_i$ + 25H$^+$ $\longrightarrow$
$$4\text{CO}_2 + 25\text{ATP} + 29\text{H}_2\text{O}$$

16. *Role of FAD as Electron Acceptor* Acyl-CoA dehydrogenase uses enzyme-bound FAD as a prosthetic group to dehydrogenate the α and β carbons of fatty acyl-CoA. What is the advantage of using FAD as an electron acceptor rather than NAD^+? Explain in terms of the standard reduction potentials for the Enz-FAD/$FADH_2$ (E'_0 = -0.219 V) and NAD^+/NADH (E'_0 = -0.320 V) half-reactions.

> *Answer* Enz-FAD, having a more positive standard reduction potential, is a better electron acceptor than NAD^+, and the reaction is driven in the direction of fatty acyl-CoA oxidation (a negative free-energy change; see Eqn 13-8). This more favorable free-energy change is obtained at the expense of 1 ATP; only 2 ATP molecules are formed per $FADH_2$ oxidized in the respiratory chain (compared with 3 ATP per NADH).

17. *β Oxidation of Arachidic Acid* How many turns of the fatty acid oxidation cycle are required to oxidize arachidic acid (see Table 9-1) completely to acetyl-CoA?

> *Answer* Arachidic acid is a 20-carbon saturated fatty acid. Nine cycles of the β-oxidation pathway are required to produce 10 molecules of acetyl-CoA, the last two in the ninth turn.

18. *Sources of H$_2$O Produced in β Oxidation* The complete oxidation of palmitate to carbon dioxide and water is represented by the overall equation

> Palmitate + $23O_2$ + $129P_i$ + 129ADP $\longrightarrow$ $16CO_2$ + 129ATP + $145H_2O$

The 145 H_2O molecules come from two separate reactions. What are they, and how many H_2O molecules are produced in each?

> *Answer* The two reactions that produce water are:
>
> 1. Palmitate + $23O_2$ $\longrightarrow$ $16CO_2$ + $16H_2O$
> 2. 129ADP + $129P_i$ $\longrightarrow$ 129ATP + $129H_2O$
>
> The palmitate oxidation reaction represents net production of water (see Problem 8). Water produced in ATP synthesis is a temporary gain because it is used up when ATP is used for work (hydrolyzed). There is no *net* production of water from ATP synthesis.

19. *Fate of Labeled Propionate* If [3-^{14}C]propionate (^{14}C in the methyl group) is added to a liver homogenate, ^{14}C-labeled oxaloacetate is rapidly produced. Draw a flow chart for the pathway by which propionate is transformed to oxaloacetate and indicate the location of the ^{14}C in oxaloacetate.

> *Answer* Propionate is first converted to the CoA derivative. Fig. 16-12 shows the three-step pathway that converts (3-C) propionyl-CoA to (4-C) succinyl-CoA, which can be summarized as:
>
> (a) Propionyl-CoA carboxylase uses Co_2 amd ATP to form D-methylmalonyl-CoA by carboxylation at C-2 of the propionyl group.
>
> (b) Methylmalonyl-CoA epimerase shifts the CoA thio-ester from C-1 (of the original propionyl group) to the newly added carboxylate, making the product L-methylmalonyl-CoA.
>
> (c) Methylmalonyl-CoA mutase moves the carboxy-CoA group from C-2 to C-3 within the original propionyl unit, forming succinyl-CoA.

Once succinyl-CoA is formed, the citric acid cycle can convert it to oxaloacetate. Based on the above reactions and the stereochemistry involved, the [^{14}C]-label is equilibrated at C-2 and C-3 of oxaloacetate.

Note: You should use these descriptions to prepare your own flow diagram.

20. *Biological Importance of Cobalt* Cattle, deer, sheep, and other ruminant animals produce large amounts of propionate in the rumen through the bacterial fermentation of ingested plant matter. This propionate is the principal source of glucose for the animals via the route

Propionate $\longrightarrow$ oxaloacetate $\longrightarrow$ glucose

In some areas of the world, notably Australia, ruminant animals sometimes show symptoms of anemia with concomitant loss of appetite and retarded growth. These symptoms are the result of the animals' inability to transform propionate to oxaloacetate, which is due to a cobalt deficiency caused by very low cobalt levels in the soil. Explain.

Answer One of the enzymes necessary for the conversion of propionate to oxaloacetate is methylmalonyl-CoA mutase (see Fig. 16-12). This enzyme requires as an essential cofactor the cobalt-containing coenzyme B_{12}, which is synthesized from vitamin B_{12}. A cobalt deficiency in animals would result in vitamin B_{12} deficiency.

CHAPTER 17 Amino Acid Oxidation and the Production of Urea

1. **Products of Amino Acid Transamination** Draw the structure and give the name of the α-keto acid resulting when the following amino acids undergo transamination with α-ketoglutarate:

   (a) Aspartate   (c) Alanine
   (b) Glutamate   (d) Phenylalanine

   Answer

   (a) $^-OOC-CH_2-\overset{\overset{O}{\|}}{C}-COO^-$ Oxaloacetate

   (c) $CH_3-\overset{\overset{O}{\|}}{C}-COO^-$ Pyruvate

   (b) $^-OOC-CH_2-CH_2-\overset{\overset{O}{\|}}{C}-COO^-$ α-Ketoglutarate

   (d) ⟨phenyl⟩$-CH_2-\overset{\overset{O}{\|}}{C}-COO^-$ Phenylpyruvate

2. **Measurement of the Alanine Aminotransferase Reaction Rate** The activity (reaction rate) of alanine aminotransferase is usually measured by including an excess of pure lactate dehydrogenase and NADH in the reaction system. The rate of alanine disappearance is equal to the rate of NADH disappearance measured spectrophotometrically. Explain how this assay works.

   Answer The measurement of the activity of alanine aminotransferase by measurement of the reaction of its product with lactate dehydrogenase is an example of a "coupled" assay. The product of the transamination (pyruvate) is rapidly consumed in the subsequent "indicator reaction," catalyzed by an excess of lactate dehydrogenase. The dehydrogenase uses the cofactor NADH, the disappearance of which is conveniently measured by observing the rate of decrease in NADH absorption at 340 nm. Thus, the rate of disappearance of NADH is a measure of the rate of the aminotransferase reaction, *if NADH and lactate dehydrogenase are added in excess.*

3. ***Distribution of Amino Nitrogen*** If your diet is rich in alanine but deficient in aspartate, will you show signs of aspartate deficiency? Explain.

> ***Answer*** No; aspartate is readily formed by the transfer of the amino group of alanine to oxaloacetate. Cellular levels of aminotransferases are sufficient to provide all of the amino acids in this fashion if the keto acids are available.

4. ***A Genetic Defect in Amino Acid Metabolism: A Case History*** A two-year-old child was brought to the hospital. His mother indicated that he vomited frequently, especially after feedings. The child's weight and physical development were below normal. His hair, although dark, contained patches of white. A urine sample treated with ferric chloride ($FeCl_3$) gave a green color characteristic of the presence of phenylpyruvate. Quantitative analysis of urine samples gave the results shown in the table below.

Substance	Concentration in patient's urine (mM)	Normal concentration in urine (mM)
Phenylalanine	7.0	0.01
Phenylpyruvate	4.8	0
Phenyllactate	10.3	0

(a) Suggest which enzyme might be deficient. Propose a treatment for this condition.

(b) Why does phenylalanine appear in the urine in large amounts?

(c) What is the source of phenylpyruvate and phenyllactate? Why does this pathway (normally not functional) come into play when the concentration of phenylalanine rises?

(d) Why does the patient's hair contain patches of white?

> ***Answer***
> (a) Since phenylalanine (and its related phenylketones) accumulate in this patient, it is likely that the first enzyme in phenylalanine catabolism, phenylalanine hydroxylase or phenylalanine-4-monoxygenase, is defective or missing (see Fig. 17-26). The most appropriate treatment for patients with this disease, known as phenylketonuria (PKU), is to establish a low-phenylalanine diet that provides just enough of the amino acid to meet the needs for protein synthesis.
>
> (b) Phenylalanine appears in the urine because high levels of this amino acid accumulate in the bloodstream and the body attempts to dispose of it.
>
> (c) Phenylalanine is converted to phenylpyruvate by transamination, a reaction that has an equilibrium constant of about 1.0. Phenyllactate is formed from phenylpyruvate by reduction (see Fig. 17-28). This pathway is of importance only when phenylalanine hydroxylase is defective.
>
> (d) The normal catabolic pathway of phenylalanine is through tyrosine, a precursor of melanin, the dark pigment normally present in hair. Decreased tyrosine levels in such patients result in varying degrees of pigment loss.

5. *Role of Cobalamin in Amino Acid Catabolism* Pernicious anemia is caused by impaired absorption of vitamin B_{12}. What is the effect of this impairment on the catabolism of amino acids? Are all amino acids affected equally? (Hint: See Box 16-2.)

> *Answer* The catabolism of the carbon skeletons of valine, isoleucine, and methionine is impaired because of the absence of a functional methylmalongyl-CoA mutase. This enzyme requires coenzyme B_{12} as a cofactor, and a deficiency of this vitamin leads to elevated methylmalonic acid levels (methylmalonic acidemia). The symptoms and effects of this deficiency are severe (see Table 17-2).

6. *Lactate versus Alanine as Metabolic Fuel: The Cost of Nitrogen Removal* The three carbons in lactate and alanine have identical states of oxidation, and animals can use either carbon source as a metabolic fuel. Compare the net ATP yield (moles of ATP per mole of substrate) for the complete oxidation (to CO_2 and H_2O) of lactate versus alanine when the cost of nitrogen excretion as urea is included.

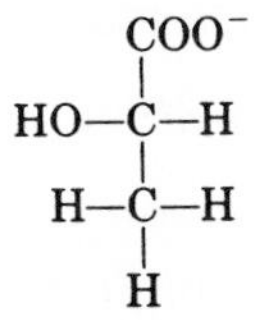

Lactate 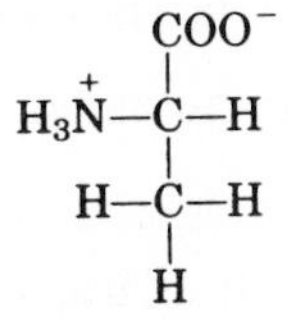   Alanine

> *Answer* Both lactate and alanine are converted to pyruvate by their respective dehydrogenases, lactate dehydrogenase and alanine dehydrogenase, producing pyruvate and $NADH + H^+$—and in the case of alanine, NH_4^+. Complete oxidation of pyruvate to CO_2 and H_2O produces 15 ATP via the citric acid cycle and oxidative phosphorylation (see Table 15-2). In addition, the NADH from each dehydrogenase reaction produces three more ATP. Thus oxidation produces 18 moles of ATP per mole of lactate.
>
> Urea formation uses the equivalent of 4 ATP per urea molecule formed (Fig. 17-11), or 2 ATP per NH_4^+. Subtracting this value from the energy yield of alanine results in 16 ATP for the oxidation of alanine.

7. *Pathway of Carbon and Nitrogen in Glutamate Metabolism* When [2-^{14}C,^{15}N]glutamate (see structure below) undergoes oxidative degradation in the liver of a rat, in which atoms of the following metabolites will each isotope be found?

(a) Urea

(b) Succinate

(c) Arginine

(d) Citrulline

(e) Ornithine

(f) Aspartate

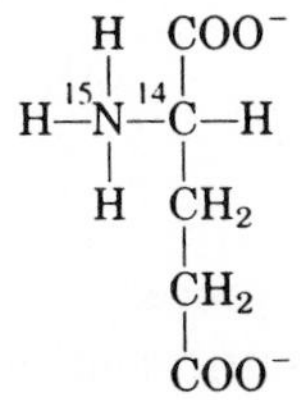

Glutamate

Answer

(a) $^{15}NH_2-CO-^{15}NH_2$

(b) $^-OO^{14}C-CH_2-CH_2-^{14}COO^-$

(c) $R-NH-\overset{\overset{\displaystyle ^{15}NH}{\|}}{C}-^{15}NH_2$

(d) $R-NH-\overset{\overset{\displaystyle O}{\|}}{C}-^{15}NH_2$

(e) No label

(f) $^-OO^{14}C-\overset{\overset{\displaystyle ^{15}NH_2}{|}}{\underset{\underset{\displaystyle H}{|}}{C}}-CH_2-^{14}COO^-$

(a)   The amino groups of urea will contain $^{15}N^-$, a result of glutamate dehydrogenase producing $^{15}NH_4^+$ or of a transaminase producing ^{15}N-labeled aspartate.

(b)   After loss of the amino group, the $[2\text{-}^{14}C]\alpha$-ketoglutarate will be metabolized in the citric acid cycle. Succinate thus formed will be labeled in the carboxyl groups.

(c)   The arginine formed in the urea cycle will contain $^{15}NH_2$- in both guanidino nitrogens.

(d)   Citrulline formed in the urea cycle will contain $^{15}NH_2$- in the carboxamide group.

(e)   No labeled N will be found in ornithine.

(f)   Aspartate will contain $^{15}NH_2$- in its amino group as a result of transamination from glutamate. It will also contain ^{14}C- in its carboxyl groups as a result of succinate (see b) conversion to oxaloacetate.

Note: In (c), (d), and (e), these intermediates in the urea cycle will contain low levels of ^{14}C as a result of a very weak synthesis of ornithine from glutamate.

8.   *Chemical Strategy of Isoleucine Catabolism* Isoleucine is degraded by a series of six steps to propionyl-CoA and acetyl-CoA:

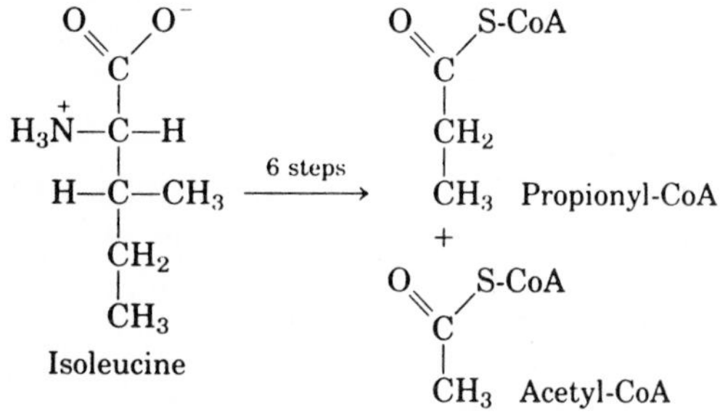

(a)   The chemical process of isoleucine degradation consists of strategies analogous to those found in the citric acid cycle and the β oxidation of fatty acids. The intermediates involved in isoleucine degradation (I to V) shown below are not in the proper order. Use your knowledge and understanding of the citric acid cycle and β-oxidation pathway to arrange the intermediates into the proper metabolic sequence for isoleucine degradation.

(b) For each step proposed above, describe the chemical process, provide an analogous example from the citric acid cycle or β-oxidation pathway, and indicate any necessary cofactors.

Answer

(a)   Isoleucine $\xrightarrow{1}$ II $\xrightarrow{2}$ IV $\xrightarrow{3}$ I $\xrightarrow{4}$ V $\xrightarrow{5}$ III $\xrightarrow{6}$ acetyl-CoA + propionyl-CoA

(b)   Step 1 is a transamination that has no analogous reaction and requires pyridoxal-P.
Step 2 is an oxidative decarboxylation similar to the pyruvate dehydrogenase reaction, and requires NAD^+.
Step 3 is an oxidation similar to the succinate dehydrogenase reaction and requires FAD.
Step 4 is a hydration, analogous to the fumarase reaction; no cofactor is required.
Step 5 is an oxidation, analogous to the malate dehyrogenase reaction of the citric acid cycle; it requires NAD^+.
Step 6 is a thiolysis, analogous to the final cleavage step of β oxidation; it requires CoA.

9. *Ammonia Intoxication Resulting from an Arginine-Deficient Diet* In a study conducted some years ago, cats were fasted overnight then given a single meal complete in amino acids but without arginine. Within 2 h, blood ammonia levels increased from a normal level of 18 μg/L to 140 μg/L, and the cats showed the clinical symptoms of ammonia toxicity. A control group fed a complete amino acid diet or an amino acid diet in which arginine was replaced by ornithine showed no unusual clinical symptoms.

(a) What was the role of fasting in the experiment?
(b) What caused the ammonia levels to rise? Why did the absence of arginine lead to ammonia toxicity? Is arginine an essential amino acid in cats? Why or why not?
(c) Why can ornithine be substituted for arginine?

Answer
(a)   The fasting resulted in lowering of blood glucose levels. Subsequent feeding of an arginine-free diet led to a rapid catabolism of all of the amino acids, especially the glucogenic ones. This catabolism was exacerbated by the lack of an essential amino acid, which prevented protein synthesis.

(b) Oxidative deamination of amino acids caused the elevation of ammonia levels. In addition, the lack of arginine (an intermediate in the urea cycle) slowed the conversion of ammonia to urea. Arginine (or ornithine) synthesis in the cat is not sufficient to meet the needs imposed by the stress of this experiment, suggesting that arginine is an essential amino acid.

(c) Ornithine (or citrulline) can be substituted for arginine because it also is an intermediate in the urea cycle.

10. *Oxidation of Glutamate* Write a series of balanced equations and the net reaction describing the oxidation of 2 mol of glutamate to 2 mol of α-ketoglutarate plus 1 mol of excreted urea.

   Answer

   H_2O + glutamate + NAD^+ $\longrightarrow$ α-ketoglutarate + NH_4^+ + NADH + H^+
   NH_4^+ + 2ATP + H_2O + CO_2 $\longrightarrow$ carbamoyl phosphate + 2ADP + P_i + $3H^+$
   Carbamoyl phosphate + ornithine $\longrightarrow$ citrulline + P_i + H^+
   Citrulline + aspartate + ATP $\longrightarrow$ argininosuccinate + AMP + PP_i + H^+
   Argininosuccinate $\longrightarrow$ arginine + fumarate
   Fumarate + H_2O $\longrightarrow$ malate
   Malate + NAD^+ $\longrightarrow$ oxaloacetate + NADH + H^+
   Oxaloacetate + glutamate $\longrightarrow$ aspartate + α-ketoglutarate
   Arginine + H_2O $\longrightarrow$ urea + ornithine

   The sum of these reactions is

   2 Glutamate + CO_2 + $4H_2O$ + $2NAD^+$ + 3ATP $\longrightarrow$
   $\qquad$ 2 α-ketoglutarate + 2NADH + $7H^+$ + urea + 2ADP + AMP + PP_i + $2P_i$ (1)

   Three additional reactions need to be considered:

   AMP + ATP $\longrightarrow$ 2ADP   (2)
   O_2 + $8H^+$ + 2NADH + 6ADP + $6P_i$ $\longrightarrow$ $2NAD^+$ + 6ATP + $8H_2O$ (3)
   H_2O + PP_i $\longrightarrow$ $2P_i$ + H^+ (4)

   Summing the last four equations:

   2 Glutamate + CO_2 + O_2 + 2ADP + $2P_i$ $\longrightarrow$ 2 α-ketoglutarate + urea + $3H_2O$ + 2ATP

11. *The Role of Pyridoxal Phosphate in Glycine Metabolism* The enzyme serine hydroxymethyl transferase (Fig. 17-23) requires a pyridoxal phosphate cofactor. Propose a mechanism for this reaction that explains the requirement. (Hint: See Fig. 17-7.)

Answer See the mechanism below.

The formaldehyde produced in the second step reacts rapidly with tetrahydrofolate at the enzyme active site to produce N^5,N^{10}-methylene-THF (see Fig. 17-19).

12. **Parallel Pathways for Amino Acid and Fatty Acid Degradation** The carbon skeleton of leucine is degraded by a series of reactions (below) closely analogous to those of the citric acid cycle and fatty acid oxidation. For each reaction, indicate its type, provide an analogous example from the citric acid cycle or β-oxidation pathways, and indicate any necessary cofactors.

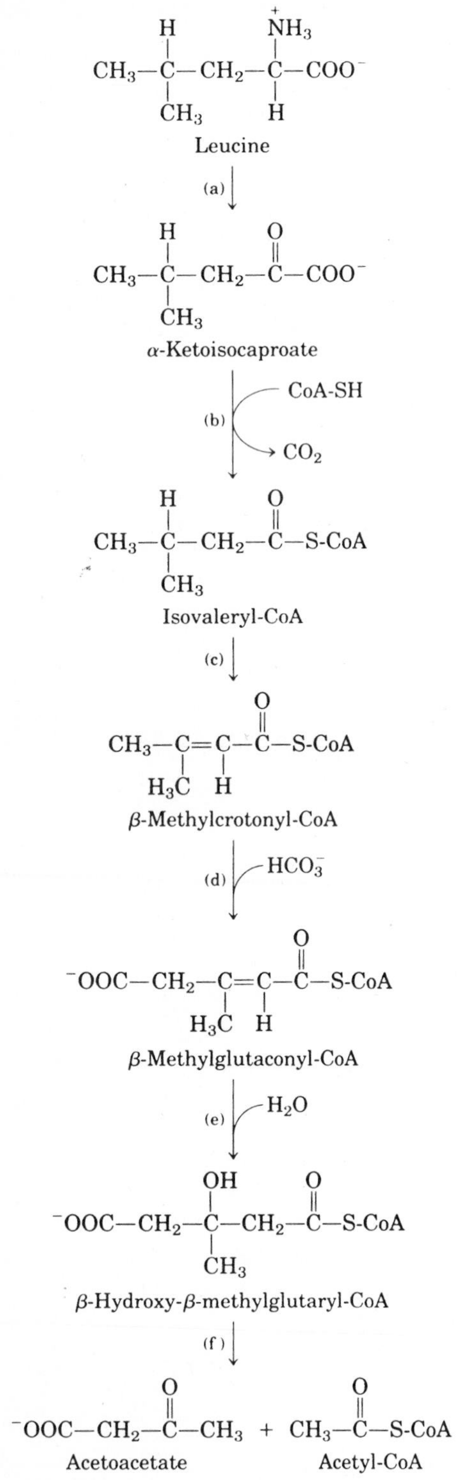

Leucine

(a)

α-Ketoisocaproate

(b)   CoA-SH → CO_2

Isovaleryl-CoA

(c)

β-Methylcrotonyl-CoA

(d) HCO_3^-

β-Methylglutaconyl-CoA

(e) H_2O

β-Hydroxy-β-methylglutaryl-CoA

(f)

Acetoacetate + Acetyl-CoA

Answer

(a) Transamination; no analogies in either pathway; requires pyridoxal-P.

(b) Oxidative decarboxylation; analogous to oxidative decarboxylation of pyruvate to acetyl-CoA prior to entry into the citric acid cycle, and of α-ketoglutarate to succinyl-CoA in the citric acid cycle; requires NAD^+, FAD, lipoate, thiamine pyrophosphate.

(c) Dehydrogenation (oxidation); analogous to dehydrogenation of succinate to fumarate in the citric acid cycle and of fatty acyl-CoA to enoyl-CoA in β oxidation; requires FAD.

(d) Carboxylation; analogous to carboxylation of pyruvate to oxaloacetate in the citric acid cycle; requires ATP and biotin.

(e) Hydration; analogous to hydration of fumarate to malate in the citric acid cycle and of enoyl-CoA to 3-hydroxyacyl-CoA in β oxidation; no cofactors.

(f) Reverse aldol reaction; analogous to reverse of citrate synthase reaction in the citric acid cycle and identical to cleavage of β-hydroxy-β-methylglutaryl-CoA in formation of ketone bodies; no cofactors.

13. *Transamination and the Urea Cycle* Aspartate aminotransferase has the highest activity of all the mammalian liver aminotransferases. Why?

Answer The second amino group introduced into urea is transferred from aspartate. This amino acid is generated in large quantities by transamination between oxaloacetate and glutamate (and many other amino acids), catalyzed by aspartate aminotransferase. Approximately one-half of all the amino groups that are excreted as urea must pass through the aspartate aminotransferase reaction, and liver contains higher levels of this aminotransferase than any other.

14. *The Case against the Liquid Protein Diet* A weight-reducing diet heavily promoted some years ago required the daily intake of "liquid protein" (soup of hydrolyzed gelatin), water, and an assortment of vitamins. All other food and drink were to be avoided. People on this diet typically lost 10 to 14 lb in the first week.

(a) Opponents argued that the weight loss was almost entirely water and would be regained almost immediately when a normal diet was resumed. What is the biochemical basis for the opponents' argument?

(b) A number of people on this diet died. What are some of the dangers inherent in the diet and how can they lead to death?

Answer

(a) A person on a diet consisting only of protein must use amino acids as the principal source of metabolic fuel. Because the catabolism of amino acids requires the removal of nitrogen as urea, the process consumes large quantities of water to dilute and excrete the urea in the urine. Furthermore, electrolytes in the "liquid protein" must be diluted with water and excreted. If this abnormally large daily water loss through the kidney is not balanced by a sufficient water intake, a net loss of body water results.

(b)   When considering the nutritional benefits of protein, keep in mind the total amount of amino acids needed for protein synthesis and the distribution of amino acids in the dietary protein. Gelatin contains a nutritionally unbalanced distribution of amino acids. As large amounts of gelatin are ingested and the excess amino acids are catabolized, the capacity of the urea cycle may be exceeded, leading to ammonia toxicity. This is further complicated by the dehydration that may result from excretion of large quantities of urea. A combination of these two factors could produce coma and death.

15. *Alanine and Glutamine in the Blood*   Blood plasma contains all the amino acids required for the synthesis of body proteins, but they are not present in equal concentrations. Two amino acids, alanine and glutamine, are present in much higher concentrations in normal human blood plasma than any of the other amino acids. Suggest possible reasons for their abundance.

Answer Muscle tissue is capable of converting amino acids to their keto acids plus ammonia and of oxidizing the keto acids to produce ATP for muscle contraction. However, urea cannot be formed in muscle. Alanine and glutamine transport amino groups to the liver (see Fig. 17-2) from muscle and other nonhepatic tissues. In muscle, amino groups from all other amino acids are transferred to pyruvate or glutamate to form alanine or glutamine, and these later amino acids are transported to the liver in the blood.

CHAPTER **18** **Oxidative Phosphorylation and Photophosphorylation**

1. *Oxidation-Reduction Reactions* The NADH dehydrogenase complex of the mitochondrial respiratory chain promotes the following series of oxidation-reduction reactions, in which Fe^{3+} and Fe^{2+} represent the iron in iron-sulfur centers, UQ is ubiquinone, UQH_2 is ubiquinol, and E is the enzyme:

(1) $NADH + H^+ + E\text{-}FMN \longrightarrow NAD^+ + E\text{-}FMNH_2$
(2) $E\text{-}FMNH_2 + 2Fe^{3+} \longrightarrow E\text{-}FMN + 2Fe^{2+} + 2H^+$
(3) $2Fe^{2+} + 2H^+ + UQ \longrightarrow 2Fe^{3+} + UQH_2$

Sum: $NADH + H^+ + UQ \rightarrow NAD^+ + UQH_2$

For each of the three reactions catalyzed by the NADH dehydrogenase complex, identify (a) the electron donor, (b) the electron acceptor, (c) the conjugate redox pair, (d) the reducing agent, and (e) the oxidizing agent.

Answer Oxidation-reduction reactions require an electron donor and an electron acceptor. Recall that electron donors are reducing agents; electron acceptors are oxidizing agents.

Reaction (1): NADH is the electron donor (a) and the reducing agent (d): E-FMN is the electron acceptor (b) and the oxidizing agent (e); $NAD^+/NADH$ and $E\text{-}FMN/E\text{-}FMNH_2$ are conjugate redox pairs (c).

Reaction (2): $E\text{-}FMNH_2$ is the electron donor (a) and reducing agent (d); Fe^{3+} is the electron acceptor (b) and oxidizing agent (e); $E\text{-}FMN/E\text{-}FMNH_2$ and Fe^{3+}/Fe^{2+} are redox pairs (c).

Reaction (3): Fe^{2+} is the electron donor (a)and reducing agent (d); UQ is the electron acceptor (b) and oxidizing agent (d); and Fe^{3+}/Fe^{2+} and UQ/UQH_2 are redox pairs (c).

2. *Standard Reduction Potentials* The standard reduction potential of any redox couple is defined for the half-cell reaction (or half-reaction):

Oxidizing agent + n electrons $\longrightarrow$ reducing agent

The standard reduction potentials of the $NAD^+/NADH$ and pyruvate/lactate redox pairs are -0.320 and -0.185 V, respectively.

(a) Which redox pair has the greater tendency to lose electrons? Explain.

(b) Which is the stronger oxidizing agent? Explain.

(c) Beginning with 1 M concentrations of each reactant and product at pH 7, in which direction will the following reaction proceed?

Pyruvate + NADH + H^+ $\rightleftharpoons$ lactate + NAD^+

(d) What is the standard free-energy change, $\Delta G^{\circ\prime}$, at 25 °C for this reaction?

(e) What is the equilibrium constant for this reaction at 25 °C?

Answer This is the same as Problem 19 of Chapter 13, except that the values of E'_0 are given in this problem to three significant figures. Recall that in oxidation-reduction reactions, electrons move from carriers of lower reduction potentials to ones of higher potentials.

(a) $NAD^+/NADH$
(b) Pyruvate/lactate
(c) Lactate + NAD^+
(d) -26.0 kJ/mol
(e) 3.59 x 10⁴

3. *Energy Span of the Respiratory Chain* Electron transfer in the mitochondrial respiratory chain may be represented by the net reaction equation

$NADH + H^+ + \frac{1}{2}O_2 \rightleftharpoons H_2O + NAD^+$

(a) Calculate the value of the change in standard reduction potential, $\Delta E'_0$, for the net reaction of mitochondrial electron transfer.

(b) Calculate the standard free-energy change, $\Delta G^{\circ\prime}$, for this reaction.

(c) How many ATP molecules could *theoretically* be generated per molecule of NADH oxidized by this reaction, given a standard free energy of ATP synthesis of 30.5 kJ/mol?

(d) How many ATP molecules could be synthesized under typical cellular conditions (see Box 13-2)?

Answer
(a), (b), and (c) These are the same as Problem 20 of Chapter 13.
(a) 1.14 V
(b) -220 kJ/mol
(c) ≈ 7
(d) Under typical cellular conditions ΔG of ATP hydrolysis is -51.8 kJ/mol (see Box 13-2); thus ΔG for ATP synthesis is 51.8 kJ/mol. The number of ATP molecules that could be generated per molecule of NADH oxidized is (220 kJ/mol)/(51.8 kJ/mol) = 4.25 ≈ 4

4. *Use of FAD Rather Than NAD$^+$ in the Oxidation of Succinate*   All the dehydrogenation steps in glycolysis and the citric acid cycle use NAD$^+$ (E'_0 for NAD$^+$/NADH = -0.32 V) as the electron acceptor except succinate dehydrogenase, which uses covalently bound FAD (E'_0 for FAD/FADH$_2$ in this enzyme = 0.05 V). Why is FAD a more appropriate electron acceptor than NAD$^+$ in the dehydrogenation of succinate? Give a possible explanation based on a comparison of the E'_0 values of the fumarate/succinate pair (E'_0 = 0.03), the NAD$^+$/NADH pair, and the succinate dehydrogenase FAD/FADH$_2$ pair.

> *Answer* From the difference in standard reduction potential ($\Delta E'_0$) for each pair of half-reactions, we can calculate the $\Delta G^{\circ\prime}$ values for the oxidation of succinate using NAD$^+$ and the oxidation using E-FAD.
> For NAD$^+$:
> $$\begin{aligned} \Delta G^{\circ\prime} &= -n\mathcal{F}\Delta E'_0 \\ &= -2(96.5 \text{ kJ/V·mol})(-0.32 \text{ V} - 0.03 \text{ V}) \\ &= 67.6 \text{ kJ/mol} \end{aligned}$$
> For E-FAD:
> $$\begin{aligned} \Delta G^{\circ\prime} &= -2(96.5 \text{ kJ/V·mol})(0.05 \text{ V} - 0.03 \text{ V}) \\ &= -3.86 \text{ kJ/mol} \end{aligned}$$
>
> The oxidation of succinate by E-FAD is favored by the negative standard free-energy change, which is consistent with a K'_{eq} of greater than 1. Oxidation by NAD$^+$ would require a large, positive, standard free-energy change and have a K'_{eq} favoring the synthesis of succinate.

5. *Degree of Reduction of Electron Carriers in the Respiratory Chain*   The degree of reduction of each electron carrier in the respiratory chain is determined by the conditions existing in the mitochondrion. For example, when the supply of NADH and O$_2$ is abundant, the steady-state degree of reduction of the carriers decreases as electrons pass from the substrate to O$_2$. When electron transfer is blocked, the carriers before the block become more reduced while those beyond the block become more oxidized (Fig. 18-7). For each of the conditions below, predict the state of oxidation of each carrier in the respiratory chain (ubiquinone and cytochromes b, c_1, c, and $a + a_3$).

(a) Abundant supply of NADH and O$_2$ but cyanide added.

(b) Abundant supply of NADH but O$_2$ exhausted.

(c) Abundant supply of O$_2$ but NADH exhausted.

(d) Abundant supply of NADH and O$_2$.

> *Answer* As shown in Figure 18-7, the oxidation-reduction state of the carriers in the electron transport system will vary with the conditions.
> (a)   Cyanide inhibits cytochrome oxidase ($a + a_3$); all carriers become reduced.
> (b)   In the absence of O$_2$, no terminal electron acceptor is present; all carriers become reduced.
> (c)   In the absence of NADH, no carrier can be reduced; all carriers become oxidized.
> (d)   These are normal circumstances (abundant NADH and O$_2$); the early carriers (UQ, for example) are somewhat reduced while the late ones (cytochrome c, for example) are oxidized.

6. *The Effect of Rotenone and Antimycin A on Electron Transfer* Rotenone, a toxic natural product from plants, strongly inhibits NADH dehydrogenase of insect and fish mitochondria. Antimycin A, a toxic antibiotic, strongly inhibits the oxidation of ubiquinol.

(a) Explain why rotenone ingestion is lethal to some insect and fish species.

(b) Explain why antimycin A is a poison.

(c) Assuming that rotenone and antimycin A are equally effective in blocking their respective sites in the electron transfer chain, which would be a more potent poison? Explain.

Answer

(a)   The inhibition of NADH dehydrogenase by rotenone decreases the rate of electron flow through the respiratory chain, which in turn decreases the rate of ATP production. If this reduced rate is unable to meet its ATP requirements, the organism dies.

(b)   Antimycin A strongly inhibits the oxidation of UQ in the respiratory chain, severely limiting the rate of electron transfer and ATP production.

(c)   Electrons flow into the system at Complex I (see Fig. 18-12) from the NAD^+-linked reactions and at Complex II from succinate and fatty acyl-CoA (see Fig. 18-9). Antimycin A inhibits electron flow (through UQ) from all of these sources whereas rotenone inhibits flow only through Complex I. Thus antimycin A is a more potent poison.

7. *Uncouplers of Oxidative Phosphorylation* In normal mitochondria the rate of electron transfer is tightly coupled to the demand for ATP. Thus when the rate of utilization of ATP is relatively low, the rate of electron transfer is also low. Conversely, when ATP is demanded at a high rate, electron transfer is rapid. Under such conditions of tight coupling, the number of ATP molecules produced per atom of oxygen consumed when NADH is the electron donor—known as the P/O ratio—is close to 3.

(a) Predict the effect of a relatively low and a relatively high concentration of an uncoupling agent on the rate of electron transfer and the P/O ratio.

(b) The ingestion of uncouplers causes profuse sweating and an increase in body temperature. Explain this phenomenon in molecular terms. What happens to the P/O ratio in the presence of uncouplers?

(c) The uncoupler 2,4-dinitrophenol was once prescribed as a weight-reducing drug. How can this agent, in principle, serve as a reducing aid? Such uncoupling agents are no longer prescribed because some deaths occurred following their use. Why can the ingestion of uncouplers lead to death?

Answer Uncouplers of oxidative phosphorylation stimulate the rate of electron flow but not ATP synthesis.

(a)   At relatively low levels of an uncoupling agent, P/O ratios will drop somewhat but the cell will be able to compensate for this by increasing the rate of electron flow; ATP levels can be kept relatively normal. At high levels of uncoupler, P/O ratios approach zero and the cell cannot maintain ATP levels.

(b)   As amounts of an uncoupler increase, the P/O ratio will decrease and the body struggles to make sufficient ATP by oxidizing more fuel. The heat produced by this increased rate of oxidation will increase the body temperature. The P/O ratio is affected as noted in (a).

(c)   Increased activity of the respiratory chain in the presence of an uncoupler requires the degradation of additional energy stores (glycogen and fat). By oxidizing more fuel in an attempt to produce the same amount of ATP, the body loses weight. If the P/O ratio nears zero, however, the lack of ATP will be lethal.

8.   *Mode of Action of Dicyclohexylcarbodiimide (DCCD)* When DCCD is added to a suspension of tightly coupled, actively respiring mitochondria, the rate of electron transfer (measured by O_2 consumption) and the rate of ATP production dramatically decrease. If a solution of 2,4-dinitrophenol is now added to the inhibited mitochondrial preparation, O_2 consumption returns to normal but ATP production remains inhibited.

(a)   What process in electron transfer or oxidative phosphorylation is affected by DCCD?

(b)   Why does DCCD affect the O_2 consumption of mitochondria? Explain the effect of 2,4-dinitrophenol on the inhibited mitochondrial preparation.

(c)   Which of the following inhibitors does DCCD most resemble in its action: antimycin A, rotenone, or oligomycin?

Answer

(a)   DCCD inhibits ATP synthesis. In tightly coupled mitochondria, this inhibition will lead to an inhibition of electron transfer also.

(b)   A decrease in electron transfer causes a decrease in O_2 consumption. 2,4-Dinitrophenol uncouples electron transfer from ATP synthesis, allowing respiration to increase. No ATP is synthesized and the P/O ratio will decrease.

(c)   Both DCCD and oligomycin inhibit ATP synthesis (see Table 18-4).

9.   *The Malate-α-Ketoglutarate Transport System of Mitochondria* The inner mitochondrial membrane transport system that promotes the transport of malate and α-ketoglutarate across the membrane (Fig. 18-25) is inhibited by *n*-butylmalonate. Suppose *n*-butylmalonate is added to an aerobic suspension of kidney cells using glucose exclusively as fuel. Predict the effect of this inhibitor on

(a) Glycolysis

(b) Oxygen consumption

(c) Lactate formation

(d) ATP synthesis

Answer NADH produced in the cytosol cannot cross the mitochondrial membrane, but must be oxidized if glycolysis is to continue. Reducing equivalents from NADH enter the mitochondrion by way of the malate-aspartate shuttle. NADH reduces oxaloacetate to form malate and NAD^+, and the malate is transported into the mitochondrion. Cytosolic oxidation of glucose can continue, and the malate is converted back to oxaloacetate and NADH in the mitochondrion. (see Fig. 18-25).

(a)   If *n*-butylmalonate, an inhibitor of the shuttle, is added to cells, NADH accumulates in the cytosol. This forces glycolysis to operate anaerobically, with reoxidation of NADH in the lactate dehydrogenase reaction.

(b)   Since reducing equivalents from the oxidation reactions of glycolysis do not enter the mitochondrion, oxygen consumption ceases.

(c)   The end product of anaerobic glycolysis, lactate, will accumulate.

(d) ATP will not be formed aerobically because the cells have converted to anaerobic glycolysis. Overall ATP synthesis will decrease drastically, to 2 ATP per glucose molecule.

10. *The Pasteur Effect* When O_2 is added to an anaerobic suspension of cells using glucose at a high rate, the rate of glucose consumption declines dramatically as the added O_2 is consumed. In addition, the accumulation of lactate ceases. This effect, first observed by Louis Pasteur in the 1860s, is characteristic of most cells capable of both aerobic and anaerobic utilization of glucose.

(a) Why does the accumulation of lactate cease after O_2 is added?

(b) Why does the presence of O_2 decrease the rate of glucose consumption?

(c) How does the onset of O_2 consumption slow down the rate of glucose consumption? Explain in terms of specific enzymes.

Answer The addition of oxygen to an anaerobic suspension allows cells to convert from fermentation to oxidative phosphorylation as a mechanism for reoxidizing NADH and making ATP. Since ATP synthesis is much more efficient under aerobic conditions, the amount of glucose needed will decrease (the Pasteur effect). This decreased utilization of glucose in the presence of oxygen can be demonstrated in any tissue that is capable of aerobic and anaerobic glycolysis.

(a) Oxygen allows the tissue to convert to oxidative phosphorylation rather than lactic acid fermentation as the mechanism for NADH oxidation.

(b) Cells produce much more ATP per glucose molecule oxidized aerobically, so less glucose is needed.

(c) As [ATP] rises in the cells, phosphofructokinase-1 is inhibited, thus slowing the rate of glucose entry into the glycolytic pathway.

11. *How Many Protons in a Mitochondrion?* Electron transfer functions to translocate protons from the mitochondrial matrix to the external medium to establish a pH gradient across the inner membrane, the outside more acidic than the inside. The tendency of protons to diffuse from the outside into the matrix, where [H^+] is lower, is the driving force for ATP synthesis via the ATP synthase. During oxidative phosphorylation by a suspension of mitochondria in a medium of pH 7.4, the internal pH of the matrix has been measured as 7.7.

(a) Calculate [H^+] in the external medium and in the matrix under these conditions.

(b) What is the outside:inside ratio of [H^+]? Comment on the energy inherent in this concentration. (Hint: See p. 383, Eqn 13-5.)

(c) Calculate the number of protons in a respiring liver mitochondrion, assuming its inner matrix compartment is a sphere of diameter 1.5 μm.

(d) From these data would you think the pH gradient alone is sufficiently great to generate ATP?

(e) If not, can you suggest how the necessary energy for synthesis of ATP arises?

Answer

(a) Using the equation pH = -log [H^+], we can calculate the external [H^+] = 4.0 x 10^{-8} M (at pH 7.4), and the internal [H^+] = 2.0 x 10^{-8} M (at pH 7.7).

(b) From (a), the ratio is 2:1. We can calculate the free energy inherent in this concentration difference across the membrane:

$$\Delta G^{\circ\prime} = RT \ln (C_2/C_1)$$
$$= (2.479 \text{ kJ/mol}) \ln 2$$
$$= -1.72 \text{ kJ/mol}$$

(c)   Given that the volume of the mitochondrion $= 4/3 \ \pi(0.75 \times 10^{-3} \text{ mm})^3$, $r = 7.5 \times 10^{-4} \text{ mm}^3$, and $[H^+] = 2.0 \times 10^{-8}$ M, the number of protons is

$$\frac{(4.19)(4.2 \times 10^{-10} \text{ mm}^3)(2.0 \times 10^{-8} \text{ mol/L})(6.023 \times 10^{23} \text{ protons/mol})}{(10^6 \text{ mm}^3/\text{L})} = 21 \text{ protons}$$

(d)   No; the total energy inherent in the pH gradient (1.72 kJ/mol—p. 559) is insufficient to synthesize 1 mol of ATP.

(e)   The overall transmembrane potential is the main factor in producing a sufficiently large ΔG (see Eqns 18-2 and 18-3).

12. ***Rate of ATP Turnover in Rat Heart Muscle***   Rat heart muscle operating aerobically fills more than 90% of its ATP needs by oxidative phosphorylation. This tissue consumes O_2 at the rate of 10 μmol/min·g of tissue, with glucose as the fuel source.

(a)   Calculate the rate at which this tissue consumes glucose and produces ATP.

(b)   If the steady-state concentration of ATP in rat heart muscle is 5 μmol/g of tissue, calculate the time required (in seconds) to completely turn over the cellular pool of ATP. What does this result indicate about the need for tight regulation of ATP production? (Note: Concentrations are expressed as micromoles per gram of muscle tissue because the tissue is mostly water.)

Answer   ATP turns over very rapidly in all types of tissues and cells.

(a)   Glucose oxidation requires 6 mol of O_2 per mol of glucose. Therefore, glucose is consumed at the rate of 10/6 = 1.7 μmol/min·g of tissue. If each glucose produces 38 ATP (see Table 15-2), the muscle produces ATP at the rate of about 1.7 x 38 = 63 $\approx$ 60 μmol/min·g, or 1 μmol/s·g.

(b)   It will take about 5 s to produce ATP at a level of 5 μmol/g. This indicates that the entire pool of ATP in muscle must be regenerated (turned over) every 5 s. In order to do this, the cell must regulate ATP synthesis very precisely.

13. ***Rate of ATP Breakdown in Flight Muscle***   ATP production in the flight muscles of the fly *Lucilia sericata* results almost exclusively from oxidative phosphorylation. During flight, 187 mL of O_2/h·g of fly body weight is needed to maintain an ATP concentration of 7 μmol/g of flight muscle. Assuming that the flight muscles represent 20% of the weight of the fly, calculate the rate at which the flight-muscle ATP pool turns over. How long would the reservoir of ATP last in the absence of oxidative phosphorylation? Assume that reducing equivalents are transferred by the glycerol-3-phosphate shuttle and that O_2 is at 25 °C and 101.3 kPa (1 atm). (Note: Concentrations are expressed in micromoles per gram of flight muscle.)

Answer   Using the gas laws (PV = nRT) we can calculate that 187 mL of O_2 contains n = PV/RT = (1 atm)(0.187 L)/(0.08205 L·atm/mol·°K)(298 °K) $\approx$ 7650 μmol of O_2.

Thus the rate of oxygen consumption by flight muscle = (7650 μmol/h)/(0.2 g)(3600 s/h) = 10.6 μmol/s·g

Assuming that ATP is formed in the ratio of 36 ATP per 6 O_2 per glucose (see Table 18-5; the glycerol-3-P shuttle produces 4 ATP from glycolysis), the amount of ATP formed is (36/6)(10.6 μmol/s·g) = 63.8 $\approx$ 64 μmol/s·g

Given a steady-state [ATP] = 7 μmol/s·g, the ATP pool would turn over every 10 s; a reservoir of 7 μmol/s·g would last 0.1 s.

14. *Transmembrane Movement of Reducing Equivalents*   Under aerobic conditions, extramitochondrial NADH must be oxidized by the mitochondrial electron transfer chain. Consider a preparation of rat hepatocytes containing mitochondria and all the enzymes of the cytosol. If [4-^{3}H]NADH is introduced, radioactivity appears quickly in the mitochondrial matrix. However, if [7-^{14}C]NADH is introduced, no radioactivity appears in the matrix. What do these observations tell us about the oxidation of extramitochondrial NADH by the electron transfer chain?

[4-^{3}H]NADH   [7-^{14}C]NADH

Answer   The malate-aspartate shuttle transfers electrons and protons from the cytoplasm into the mitochondrion. Neither NAD$^+$ nor NADH passes through the inner membrane, thus the labeled NAD moiety of [7-^{14}C]NADH remains in the cytosol. The ^{3}H on [4-^{3}H]NADH enters the mitochondrion via the malate-aspartate shuttle (see Fig 18-25). In the cytosol, [4-^{3}H]NADH transfers its ^{3}H to oxaloacetate to form [^{3}H]malate, which is transported into the mitochondrion. The unlabeled NAD$^+$ that is produced remains in the cytosol. In the mitochondrion, [^{3}H]malate donates the ^{3}H to NAD$^+$ to form [4-^{3}H]NADH.

15. *Photochemical Efficiency of Light at Different Wavelengths*   The rate of photosynthesis, measured by O$_2$ production, is higher when a green plant is illuminated with light of wavelength 680 nm than with light of 700 nm. However, illumination by a combination of light of 680 nm and 700 nm gives a higher rate of photosynthesis than light of either wavelength alone. Explain.

Answer   There are two photosystems that drive photosynthesis and photophosphorylation in plants. Photosystem I absorbs light maximally at 700 nm and catalyzes cyclic photophosphorylation and NADP$^+$ reduction (see Fig. 18-44). Photosystem II absorbs light maximally at 680 nm, splits H$_2$O to O$_2$ and H$^+$, and donates electrons and H$^+$ to Photosystem I. Therefore, light of 680 nm is better in promoting O$_2$ production, but maximum photosynthetic rates are observed only when plants are illuminated with light of both wavelengths.

16. **Role of H_2S in Some Photosynthetic Bacteria** Illuminated purple sulfur bacteria carry out photosynthesis in the presence of H_2O and $^{14}CO_2$, but only if H_2S is added and O_2 is absent. During the course of photosynthesis, measured by formation of [^{14}C]carbohydrate, H_2S is converted into elemental sulfur, but no O_2 is evolved. What is the role of the conversion of H_2S into sulfur? Why is no O_2 evolved?

> **Answer** Purple sulfur bacteria use H_2S as a source of electrons and protons
>
> $$H_2S \longrightarrow S + 2H^+ + 2e^-$$
>
> The electrons are "activated" by a light energy-capturing photosystem. These cells produce their ATP by photophosphorylation and their NADPH from H_2S oxidation. Since H_2O is not split, O_2 is not evolved. Photosystem II is absent in these bacteria, further explaining why O_2 is not evolved.

17. **Boosting the Reducing Power of Photosystem I by Light Absorption** When photosystem I absorbs red light at 700 nm, the standard reduction potential of P700 changes from 0.4 to about -1.2 V. What fraction of the absorbed light is trapped in the form of reducing power?

> **Answer** For a change in standard reduction potential of 0.4 to -1.2 V, the free-energy change per electron is
>
> $$\begin{aligned} \Delta G^{\circ\prime} &= n\mathcal{F}\Delta E \\ &= -(96.5 \text{ kJ/V·mol})(-1.6 \text{ V}) \\ &= 154 \text{ kJ/mol} \end{aligned}$$
>
> Two photons are absorbed per electron elevated to a higher energy level, which for 700 nm light is equivalent to 2(170 kJ/mol) = 340 kJ/mol (see Fig. 18-35). Thus the fraction of light energy trapped as reducing power is
>
> $$(154 \text{ kJ/mol})/(340 \text{ kJ/mol}) = 0.45$$

18. **Mode of Action of the Herbicide DCMU** When chloroplasts are treated with 3-(3,4-dichlorophenyl)-1,1-dimethylurea (DCMU, or Diuron), a potent herbicide, O_2 evolution and photophosphorylation cease. Oxygen evolution but not photophosphorylation can be restored by the addition of an external electron acceptor, or Hill reagent. How does this herbicide act as a weed killer? Suggest a location for the inhibitory site of this herbicide in the scheme shown in Figure 18-44. Explain.

> **Answer** DCMU must inhibit the electron transfer system linking Photosystem II and Photosystem I at a position ahead of the first site of ATP production. DCMU competes with Q_B for electrons from Q_A (see Table 18-4). Thus, addition of a Hill reagent allows H_2O to be split and O_2 to be evolved, but electrons are pulled out of the system before the point of ATP synthesis and before the production of NADPH. DCMU kills plants by inhibiting ATP production.

19. **Bioenergetics of Photophosphorylation** The steady-state concentrations of ATP, ADP, and P_i in isolated spinach chloroplasts under full illumination at pH 7.0 are 120, 6, and 700 μM, respectively.

   (a) What is the free-energy requirement for the synthesis of 1 mol of ATP under these conditions?

(b) The energy for ATP synthesis is furnished by light-induced electron transfer in the chloroplasts. What is the minimum voltage drop necessary during the transfer of a pair of electrons to synthesize ATP under these conditions? (You may need to refer to p. 389, Eqn 13-8.)

Answer

(a) $\Delta G = \Delta G^{\circ\prime} + RT \ln \dfrac{[ATP]}{[ADP][P_i]}$

$= 30.5 \text{ kJ/mol} + (2.479 \text{ kJ/mol}) \ln \dfrac{1.2 \times 10^{-4} \text{ M}}{(6 \times 10^{-6} \text{ M})(7 \times 10^{-4} \text{ M})}$

$= 30.5 \text{ kJ/mol} + 25.4 \text{ kJ/mol}$

$= 55.9 \text{ kJ/mol} \approx 56 \text{ kJ/mol}$

(b) $\Delta G = -n\mathscr{F}\Delta E$

$\Delta E = -\Delta G/n\mathscr{F}$

$= \dfrac{-(55.9 \text{ kJ/mol})}{2(96.5 \text{ kJ/V·mol})}$

$= 0.29 \text{ V}$

20. *Equilibrium Constant for Water-Splitting Reactions* The coenzyme $NADP^+$ is the terminal electron acceptor in chloroplasts, according to the reaction

$$2H_2O + 2NADP^+ \longrightarrow 2NADPH + 2H^+ + O_2$$

Use the information in Table 18-2 to calculate the equilibrium constant at 25 °C for this reaction. (The relationship between K'_{eq} and $\Delta G^{\circ\prime}$ is discussed on p. 368.) How can the chloroplast overcome this unfavorable equilibrium?

Answer Using standard reduction potentials from Table 18-2, $\Delta E'_0$ for the reaction is -0.324 V - 0.816 V = -1.14 V.

$\Delta G^{\circ\prime} = -n\mathscr{F}\Delta E'_0$

$= -4(96.5 \text{ kJ/V·mol})(-1.14 \text{ V})$

$= 440 \text{ kJ/mol}$

(Note that $n = 4$ because 4 electrons are required to produce 1 mol of O_2.)

$\Delta G^{\circ\prime} = -RT \ln K'_{eq}$

$\ln K'_{eq} = -\Delta G^{\circ\prime}/RT$

$= (-440 \text{ kJ/mol})/(2.479 \text{ kJ/mol})$

$= -177.5$

$K'_{eq} = 1.2 \times 10^{-77}$

The equilibrium is clearly very unfavorable. In chloroplasts, the input of light energy overcomes this barrier.

21. *Energetics of Phototransduction* During photosynthesis, eight photons of light must be absorbed (four by each photosystem) for every O_2 molecule produced:

$$2H_2O + 2NADP^+ + 8 \text{ photons} \longrightarrow 2NADPH + 2H^+ + O_2$$

Assuming that these photons have a wavelength of 700 nm (red) and that the absorption and utilization of light energy are 100% efficient, calculate the free-energy change for the process.

Answer From Problem 20, $\Delta G^{o\prime}$ for the production of 1 mol of O_2 is 440 kJ/mol. A light input of 8 photons (700 nm) is equivalent to 8(170 kJ/mol) = 1360 kJ/mol (see Fig. 18-25). Thus coupling these two processes, the overall standard free-energy change is

$\Delta G^{o\prime}$ = (440 - 1360) kJ/mol = -920 kJ/mol

22. ***Electron Transfer to a Hill Reagent*** Isolated spinach chloroplasts evolve O_2 when illuminated in the presence of potassium ferricyanide (the Hill reagent), according to the equation

$$2H_2O + 4Fe^{3+} \longrightarrow O_2 + 4H^+ + 4Fe^{2+}$$

where Fe^{3+} represents ferricyanide and Fe^{2+}, ferrocyanide. Is NADPH produced in this process? Explain.

Answer No NADPH is produced. Artificial electron acceptors can remove electrons from the photosynthetic apparatus and stimulate O_2 production. Ferricyanide competes with the cytochrome *bf* complex for electrons and removes them from the system. Consequently, P700 does not receive any electrons that can be activated for $NADP^+$ reduction. However, O_2 is evolved because all components of Photosystem II are oxidized (see Fig. 18-44).

23. ***How Often Does a Chlorophyll Molecule Absorb a Photon?*** The amount of chlorophyll *a* (M_r 892) in a spinach leaf is about 20 μg/cm^2 of leaf. In noonday sunlight (average energy 5.4 J/cm^2·min), the leaf absorbs about 50% of the radiation. How often does a single chlorophyll molecule absorb a photon? If the average lifetime of an excited chlorophyll molecule in vivo is 1 ns, what fraction of chlorophyll molecules are excited at any one time?

Answer The leaf absorbs light in units of photons that vary in energy between 170 and 300 kJ/mol, depending on wavelength (see p. 574). The leaf absorbs light energy at the rate of 0.5(5.4 J/cm^2·min) = 2.7 J/cm^2·min. Assuming an average energy of 270 kJ/mol of photons, this rate of light absorption is

$$(2.7 \times 10^{-3} \text{ kJ/cm}^2\text{·min})/(270 \text{ kJ/mol photons}) = 1 \times 10^{-5} \text{ mol/cm}^2\text{·min}$$

The concentration of chlorophyll in the leaf is

$$(20 \times 10^{-6} \text{ g/cm}^2)/(892 \text{ g/mol}) = 2.24 \times 10^{-8} \text{ mol/cm}^2$$

Thus, 1 mol or 1 molecule of chlorophyll absorbs 1 mol of photons every
$$\begin{aligned}(2.24 \times 10^{-8} \text{ mol/cm}^2)/(1 \times 10^{-5} \text{ mol/cm}^2\text{·min}) &= 2.24 \times 10^{-3} \text{ min} \\ &= 0.13 \text{ s} \\ &\approx 1 \times 10^{-1} \text{ s or 100 ms.}\end{aligned}$$

Since excitation lasts about 1×10^{-9} s, the fraction of chlorophylls excited at any one time is $(1 \times 10^{-9} \text{ s})/(0.1 \text{ s}) = 1 \times 10^{-8}$, or one in every 10^8 molecules.

24. ***Effect of Monochromatic Light on Electron Flow*** The extent to which an electron carrier is oxidized or reduced during photosynthetic electron transfer can sometimes be observed directly with a spectrophotometer. When chloroplasts are illuminated with 700 nm light, cytochrome *f*, plastocyanin, and plastoquinone are oxidized. When chloroplasts are illuminated with 680 nm light, however, these electron carriers are reduced. Explain.

Answer Light at 700 nm activates electrons in P700 and $NADP^+$ is reduced (see Fig. 18-44). This drains all of the electrons from the electron transport system between Photosystems II and I because light at 680 nm is not available to replace electrons by activating Photosystem II. When light at 680 nm activates Photosystem II (but not Photosystem I), all of the carriers between the two systems become reduced because no electrons can be excited in Photosystem I.

25. *Function of Cyclic Photophosphorylation* When the $[NADPH]/[NADP^+]$ ratio in chloroplasts is high, photophosphorylation is predominantly cyclic (Fig. 18-44). Is O_2 evolved during cyclic photophosphorylation? Explain. Can the chloroplast produce NADPH this way? What is the main function of cyclic photophosphorylation?

Answer At high $[NADPH]/[NADP^+]$ ratios, electron transport from reduced ferredoxin to $NADP^+$ is inhibited and the electrons are diverted into the cytochrome *bf* complex. These electrons return to P700 and ATP is synthesized by photophosphorylation. Since electrons are not lost from P700, none are needed from Photosystem II and H_2O is not split to produce O_2. In addition, NADPH is not formed because the electrons return to P700. The function of cyclic photophosphorylation is to produce ATP.

CHAPTER 19

Carbohydrate Biosynthesis

1. ***Role of Oxidative Phosphorylation in Gluconeogenesis*** Is it possible to obtain a net synthesis of glucose from pyruvate if the citric acid cycle and oxidative phosphorylation are totally inhibited?

   Answer No; the transformation of two molecules of pyruvate to one molecule of glucose requires an input of energy (4 ATP + 2 GTP) and reducing power (2 NADPH), which are obtained only through the citric acid cycle and oxidative phosphorylation pathway by catabolism of amino acids, fatty acids, or other carbohydrates.

2. ***Pathway of Atoms in Gluconeogenesis*** A liver extract capable of carrying out all the normal metabolic reactions of the liver is briefly incubated in separate experiments with the ^{14}C-labeled precursors shown below. Trace the pathway of each precursor through gluconeogenesis. Indicate the location of ^{14}C in all intermediates of the process and in the product, glucose.

   (a) [^{14}C]Bicarbonate, HO—^{14}C(=O)—O$^-$

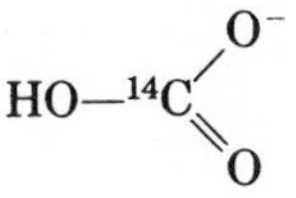

   (b) [1-^{14}C]Pyruvate, CH$_3$—C(=O)—^{14}COO$^-$

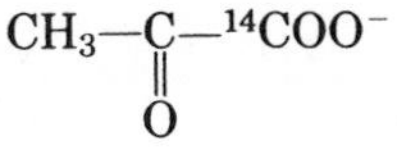

   Answer
   (a) In the pyruvate carboxylase reaction, ^{14}CO$_2$ is added to pyruvate to form [4-^{14}C] oxaloacetate, but the phosphoenolpyruvate carboxykinase reaction removes the *same* CO$_2$ in the next step. Thus ^{14}C is not (initially) incorporated into glucose.

(b)

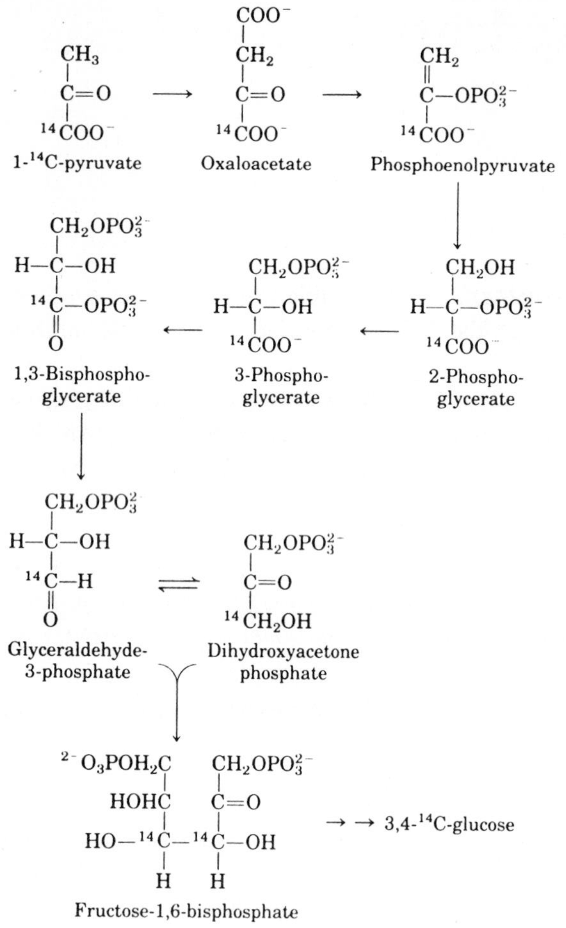

$$
\begin{array}{ccccc}
& & \text{COO}^- & & \\
& & | & & \\
\text{CH}_3 & & \text{CH}_2 & & \text{CH}_2 \\
| & & | & & \parallel \\
\text{C}=\text{O} & \longrightarrow & \text{C}=\text{O} & \longrightarrow & \text{C}-\text{OPO}_3^{2-} \\
| & & | & & | \\
{}^{14}\text{COO}^- & & {}^{14}\text{COO}^- & & {}^{14}\text{COO}^- \\
\text{1-}{}^{14}\text{C-pyruvate} & & \text{Oxaloacetate} & & \text{Phosphoenolpyruvate}
\end{array}
$$

1,3-Bisphospho-glycerate ← 3-Phospho-glycerate ← 2-Phospho-glycerate

Glyceraldehyde-3-phosphate ⇌ Dihydroxyacetone phosphate

→ → 3,4-^{14}C-glucose

Fructose-1,6-bisphosphate

3. ***Pathway of CO$_2$ in Gluconeogenesis*** In the first bypass step in gluconeogenesis, the conversion of pyruvate to phosphoenolpyruvate, pyruvate is carboxylated by pyruvate carboxylase to oxaloacetate and is subsequently decarboxylated by PEP carboxykinase to yield phosphoenolpyruvate. The observation that the addition of CO$_2$ is directly followed by the loss of CO$_2$ suggests that ^{14}C of ^{14}CO$_2$ would not be incorporated into PEP, glucose, or any of the intermediates in gluconeogenesis. However, it has been found that if rat liver slices synthesize glucose in the presence of ^{14}CO$_2$, ^{14}C slowly appears in PEP and eventually appears in C-3 and C-4 of glucose. How does the ^{14}C-label get into PEP and glucose? (Hint: During gluconeogenesis in the presence of ^{14}CO$_2$, several of the four-carbon citric acid cycle intermediates also become labeled.)

> ***Answer*** Because pyruvate carboxylase is a mitochondrial enzyme, the [^{14}C]oxaloacetate (OAA) formed by this reaction mixes with the OAA pool of the citric acid cycle. A mixture of [1-^{14}C] and [4-^{14}C]OAA forms by randomization of the C-1 and C-4 positions in the reversible conversions of OAA to malate to succinate. [1-^{14}C]OAA leads to formation of [3,4-^{14}C]glucose.

4. **Regulation of Fructose-1,6-Bisphosphatase and Phosphofructokinase-1** What are the effects of increasing concentrations of ATP and AMP on the catalytic activities of fructose-1,6-bisphosphatase and phosphofructokinase-1? What are the consequences of these effects on the relative flow of metabolites through gluconeogenesis and glycolysis?

Answer The two enzymes are regulated by adenine nucleotide modifiers in a reciprocal manner: what inhibits one activates the other. PFK-1, which catalyzes the rate-limiting step in glycolysis, is activated by AMP but inhibited by ATP. FBPase, which catalyzes the reverse reaction by hydrolysis, is activated by ATP and inhibited by AMP. This relates to "adenylate energy charge": when there is a need for energy (ATP), the flow of metabolites through glycolysis is stimulated and gluconeogenesis is inhibited. Conversely, when there is an abundance of energy, glycolysis is inhibited and the flow in the direction of gluconeogenesis is stimulated.

5. **Glucogenic Substrates** A common procedure for determining the effectiveness of compounds as precursors of glucose in mammals is to fast the animal until the liver glycogen stores are depleted and then administer the substrate in question. A substrate that leads to a *net* increase in liver glycogen is termed glucogenic because it must first be converted to glucose-6-phosphate. Show by means of known enzymatic reactions which substrates are glucogenic:

(a) $^-OOC-CH_2-CH_2-COO^-$
  Succinate

(b) $\quad$ OH $\quad$ OH $\quad$ OH
  $\quad\quad |\quad\quad |\quad\quad |$
  CH_2-C-CH_2
  $\quad\quad\quad |$
  $\quad\quad\quad$ H
  Glycerol

(c) $\quad\quad\quad$ O
  $\quad\quad\quad ||$
  $CH_3-C-S-CoA$
  Acetyl-CoA

(d) $\quad\quad\quad$ O
  $\quad\quad\quad ||$
  $CH_3-C-COO^-$
  Pyruvate

(e) $\quad CH_3-CH_2-CH_2-COO^-$
  Butyrate

Answer
(a) Succinate is converted by the citric acid cycle to fumarate by succinate dehydrogenase, then to malate by fumarase, then to oxaloacetate by malate dehydrogenase. OAA can then leave the mitochondria via the malate-aspartate shuttle, and is converted to PEP, which is glucogenic in the cytosol.

(b) Glycerol is converted to glycerol-1-P by glycerol kinase, then by a dehydrogenase (using NAD^+) to dihydroxyacetone-P, which is glucogenic.

(c)   Acetyl-CoA is not glucogenic. Higher animals do not have the enzymes to convert it to pyruvate.

(d)   Pyruvate is converted to oxaloacetate by pyruvate carboxylase, which is used for gluconeogenesis as in (a).

(e)   Butyrate is converted to butyryl-CoA by an acyl-CoA synthetase, and a single turn of the β-oxidation pathway converts butyryl-CoA to two molecules of acetyl-CoA, which is not glucogenic.

6.   ***Blood Lactate Levels during Vigorous Exercise*** The concentration of lactate in blood plasma before, during, and after a 400 m sprint are shown below.

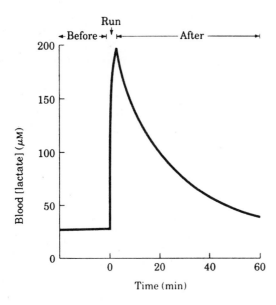

(a)   What causes the rapid rise in lactate concentration?

(b)   What causes the decline in lactate concentration after completion of the run? Why does the decline occur more slowly than the increase?

(c)   Why is the concentration of lactate not zero during the resting state?

Answer
(a)   Due to rapid depletion of ATP during strenuous muscular exertion, the rate of glycolysis increases dramatically, producing increased cytosolic concentrations of pyruvate and NADH, which are converted to lactate and NAD^+ by lactate dehydrogenase (lactic acid fermentation).

(b)   When energy demands are reduced, the oxidative capacity of the mitochondria is again adequate, and lactate is transformed to pyruvate by lactase dehydrogenase and thus to glucose. The rate of the dehydrogenase reaction is slower in this direction because of the limited availability of NAD^+ and because the equilibrium of the reaction is strongly in favor of lactate (conversion of lactate to pyruvate is energy-requiring).

(c)   The equilibrium of the lactate dehydrogenase reaction

$$Puruvate + NADH + H^+ \rightleftharpoons lactate + NAD^+$$

is *strongly* in favor of lactate. Thus, even at very low concentrations of NADH and pyruvate, there is a significant concentration of lactate.

7. ***Excess O_2 Uptake during Gluconeogenesis*** Lactate absorbed by the liver is converted to glucose. This process requires the input of 6 mol of ATP for every mole of glucose produced. The extent of this process in rat liver slices can be monitored by administering [^{14}C]lactate and measuring the amount of [^{14}C]glucose produced. Because the stoichiometry of O_2 consumption and ATP production is known (Chapter 18), we can predict the extra O_2 consumption above the normal rate when a given amount of lactate is administered. The extra amount of O_2 necessary for the synthesis of glucose from lactate, however, when actually measured is always higher than predicted by known stoichiometric relationships. Suggest a possible explanation for this observation.

   Answer If the catabolic and biosynthetic pathways are operating simultaneously, a certain amount of ATP is consumed via futile cycles, in which no useful work is done. Examples of such futile cycles (which are usually blocked by reciprocal regulation of kinases and hydrolases) are between glucose and glucose-6-P or between fructose-6-P and fructose-1,6-bisphosphate. The net hydrolysis of ATP to ADP and P_i increases the consumption of oxygen, the terminal electron acceptor in oxidative phosphorylation.

8. ***At What Point Is Glycogen Synthesis Regulated?*** Explain how the two following observations identify the point of regulation in the synthesis of glycogen in skeletal muscle:

   (a) The measured activity of glycogen synthase in resting muscle, expressed in micromoles of UDP-glucose used per gram per minute, is lower than the activity of phosphoglucomutase or UDP-glucose pyrophosphorylase, each measured in terms of micromoles of substrate transformed per gram per minute.

   (b) Stimulation of glycogen synthesis leads to a small decrease in the concentrations of glucose-6-phosphate and glucose-1-phosphate, a large decrease in the concentration of UDP-glucose, but a substantial increase in the concentration of UDP.

   Answer The observation (a) that glycogen synthase has the lowest measured activity of the enzymes involved in glycogen synthesis suggests that this step is the kinetic bottleneck in the flow of metabolites and is thus a prime candidate for regulation. This is confirmed by the observation (b) that *stimulation* of glycogen synthesis by activation of the regulatory enzyme leads to a *decrease* in the steady-state concentrations of the intermediates *before* the regulation point (especially UDP-glucose) and an *increase* in the concentration of metabolites *after* the regulation point (UDP).

9. ***What is the Cost of Storing Glucose as Glycogen?*** Write the sequence of steps and the net reaction required to calculate the cost in number of ATPs of converting cytosolic glucose-6-phosphate into glycogen and back into glucose-6-phosphate. What fraction of the maximum number of ATPs that are available from complete catabolism of glucose-6-phosphate to CO_2 and H_2O does this cost represent?

   Answer For synthesis of glycogen:
   Glucose-6-P $\longrightarrow$ glucose-1-P
   Glucose-1-P + UTP $\longrightarrow$ UDP-glucose + PP_i
   PP_i + H_2O $\longrightarrow$ 2P_i
   UDP-glucose + glycogen$_n$ $\longrightarrow$ UDP + glycogen$_{n+1}$
   UDP + ATP $\longrightarrow$ UTP + ADP

Thus, the net reaction for synthesis is:

Glucose-6-P + ATP + H_2O + glycogen$_n$ $\longrightarrow$ glycogen$_{n+1}$ + ADP + $2P_i$

For breakdown of glycogen:

Glycogen$_{n+1}$ + P_i $\longrightarrow$ glucose-1-P + glycogen$_n$
Glucose-1-P $\longrightarrow$ glucose-6-P

Thus, the net reaction for breakdown is:

Glycogen$_{n+1}$ + P_i $\longrightarrow$ glucose-6-P + glycogen$_n$

The net reaction for synthesis and breakdown:

ATP + H_2O $\longrightarrow$ ADP + P_i

Each mole of glucose-6-P converted to glycogen and back to glucose-6-P costs 1 ATP. Each mole of glucose-6-P completely catabolized yields 39 mol of ATP (see Table 15-2). Thus, the fraction is $1/39 = 0.026$, or 2.6%. The efficiency of energy storage is 97.4%.

10. *Identification of a Defective Enzyme in Carbohydrate Metabolism* A sample of liver tissue was obtained post mortem from the body of a patient believed to be genetically deficient in one of the enzymes of carbohydrate metabolism. A homogenate of the liver sample had the following characteristics: (1) it degraded glycogen to glucose-6-phosphate, (2) it was unable to make glycogen from any sugar or to utilize galactose as an energy source, and (3) it synthesized glucose-6-phosphate from lactate. Which of the following three enzymes was deficient?

(a) Glycogen phosphorylase

(b) Fructose-1,6-bisphophatase

(c) UDP-glucose pyrophosphorylase

Give reasons for your choice.

> *Answer* Characteristic (1) indicates that glycogen phosphorylase is functioning, and (2) indicates that fructose-1,6-bisphosphatase is operative. Thus the deficiency is in **(c)** UDP-glucose pyrophosphorylase. This enzyme is involved in the synthesis of UDP-glucose, a key intermediate in glycogen synthesis and in conversion of galactose (by epimerization) to glucose, a reaction required for the use of galactose as an energy source. Without this enzyme, starting from lactate the reversal of glycolysis would stop at formation of glucose-6-P.

11. *Ketosis in Sheep* The udder of a ewe uses almost 80% of the total glucose synthesized by the animal. The glucose is used for milk production, principally in the synthesis of lactose and of glycerol-3-phosphate, used in the formation of milk triacylglycerols. During the winter when food quality is poor, milk production decreases and the ewes sometimes develop ketosis, that is, increased levels of plasma ketone bodies. Why do these changes occur? A standard treatment for this condition is the administration of large doses of propionate (which is readily converted to succinyl-CoA in ruminants). How does this treatment work?

Answer   In ruminants such as sheep, ingested food is fermented by bacteria to yield acetate, lactate, and propionate. In the ewe, the latter two substances are transformed to glucose via gluconeogenesis and are subsequently used for milk production. In the absence of an adequate food supply, the ewe must use glucogenic amino acids from body protein to produce glucose for limited milk production. The energy needs of the ewe are met by the catabolism of body fat and ketogenic amino acids. The diversion of glucose and its precursor oxaloacetate to milk production under conditions of extensive fatty acid catabolism results in ketosis and associated acidosis. This situation is analogous to that in a diabetic. Ruminants can readily transform propionate to succinyl-CoA (via the intermediates propionyl-CoA, D-methylmalonyl-CoA, and L-methylmalonyl-CoA) and thus to oxaloacetate. In the presence of propionate the activity of the citric acid cycle is therefore increased, and ketosis is averted.

12. *Adaptation to Galactosemia*   Galactosemia is a pathological condition in which there is deficient utilization of galactose derived from lactose in the diet. One form of this disease is due to the absence of the enzyme UDP-glucose:galactose-1-phosphate uridylyltransferase. If an individual survives the disease in early life, some capacity to metabolize ingested galactose may develop in later life, because of increased production of the enzyme UDP-galactose pyrophosphorylase, which catalyzes the reaction

$$\text{Galactose-1-phosphate} + \text{UTP} \longrightarrow \text{UDP-galactose} + \text{PP}_i$$

How does the presence of this enzyme increase the capacity of such individuals to metabolize galactose?

Answer   UDP-galactose pyrophosphorylase allows the catabolism of galactose by the following route:

$$\text{Galactose} + \text{ATP} \overset{1}{\longrightarrow} \text{galactose-1-P} + \text{ADP}$$

$$\text{Galactose-1-P} + \text{UTP} \overset{2}{\longrightarrow} \text{UDP-galactose} + \text{PP}_i$$

$$\text{UDP-galactose} \overset{3}{\longrightarrow} \text{UDP-glucose}$$

The net reaction:

$$\text{Galactose} + \text{ATP} + \text{UTP} \longrightarrow \text{UDP-glucose} + \text{ADP} + \text{PP}_i$$

Step 1 is catalyzed by galactokinase, step 2 by UDP-galactose pyrophosphorylase, and step 3 by galactose-4-epimerase. UDP-glucose is subsequently used to synthesize glycogen or is hydrolyzed to UMP and glucose-1-P.

13. *Phases of Photosynthesis*   When a suspension of green algae is illuminated in the absence of CO_2 and then incubated with $^{14}CO_2$ in the dark, $^{14}CO_2$ is converted into [^{14}C]glucose for a brief time. What is the significance of this observation with regard to the CO_2 fixation process and how is it related to the light reactions of photosynthesis? Why does the conversion of $^{14}CO_2$ into [^{14}C]glucose stop after a brief time?

> *Answer*   This observation suggests that photosynthesis occurs in two phases: (1) a light-dependent phase that generates ATP and NADPH, which are essential for CO_2 fixation, and (2) a light-independent (dark) phase in which these energy-rich components are used for synthesis of glucose. In the absence of additional illumination, the supplies of NADPH and ATP become exhausted and CO_2 fixation ceases.

14. *Identification of Key Intermediates in CO_2 Fixation*   Calvin and his colleagues used the unicellular green alga *Chlorella* to study the carbon fixation reactions of photosynthesis. In their experiments $^{14}CO_2$ was incubated with illuminated suspensions of algae under different conditions. They followed the time course of appearance of ^{14}C in two compounds, X and Y, under two sets of conditions.

(a) Illuminated *Chlorella* were grown on unlabeled CO_2; then the lights were turned off, and $^{14}CO_2$ was added (vertical dashed line in graph **a**). Under these conditions X was the first compound to become labeled with ^{14}C. Compound Y was unlabeled.

(b) Illuminated *Chlorella* cells were grown in $^{14}CO_2$. Illumination was continued until all the $^{14}CO_2$ had disappeared (vertical dashed line in graph **b**). Under these conditions compound X became labeled quickly but lost its radioactivity with time, whereas compound Y became more radioactive with time.

Suggest the identities of X and Y based on your understanding of the Calvin cycle.

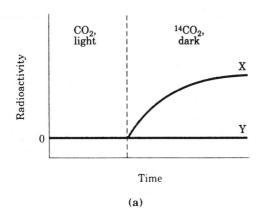

(a)

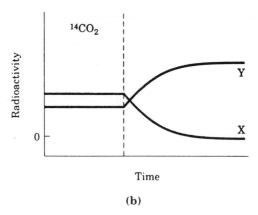

(b)

Answer Compound X is 3-phosphoglycerate and compound Y is ribulose-1,5-bisphosphate (see Fig. 19-19).

(a) Illumination of *Chlorella* in the presence of unlabeled CO_2 gives rise to steady-state levels of ribulose-1,5-bisphosphate, 3-phosphoglycerate, ATP, and NADPH. When the light is turned off the production of ATP and NADPH ceases, but the Calvin cycle continues briefly until the residual ATP and NADPH are exhausted. Once this occurs, the conversion of 3-phosphoglycerate (Stage 2 in Fig. 19-19) to hexoses, which depends on ATP and NADPH, is blocked. Thus the $^{14}CO_2$ added at the time the light is turned off is transformed primarily to 3-phosphoglycerate, but not to other intermediates of the cycle, such as ribulose-1,5-bisphosphate.

(b) Illumination of *Chlorella* in the presence of $^{14}CO_2$ gives rise to steady-state levels of ^{14}C-labeled 3-phosphoglycerate and ribulose-1,5-bisphosphate. If the concentration of CO_2 is rapidly decreased, none is available for the ribulose-1,5-bisphosphate carboxylase reaction, which constitutes a block at the fixation step (Stage 1 in Fig. 19-19). Since this experiment is carried out under conditions of constant illumination, the steps requiring ATP and NADPH are not blocked and all labeled 3-phosphoglycerate (X) can be converted to labeled ribulose-1,5-bisphosphate (Y). This results in a decrease in labeled X and a commensurate increase in labeled Y.

15. ***Pathway of CO_2 Fixation in Maize*** If a maize (corn) plant is illuminated in the presence of $^{14}CO_2$, after about 1 s more than 90% of all the radioactivity incorporated in the leaves is found in the C-4 atoms of malate, aspartate, and oxaloacetate. Only after 60 s does ^{14}C appear in the C-1 atom of 3-phosphoglycerate. Explain.

Answer In maize, CO_2 is fixed by the C_4 pathway worked out by Hatch and Slack. Phosphoenolpyruvate is rapidly carboxylated to oxaloacetate, some of which undergoes transamination to aspartate but most of which is reduced to malate in the mesophyll cells. Only after subsequent decarboxylation of labeled malate does $^{14}CO_2$ enter the Calvin cycle for conversion to glucose. The rate of entry into the cycle is limited by the rate of the rubisco-catalyzed reaction.

16. ***Chemistry of Malic Enzyme: Variation on a Theme*** Malic enzyme, found in the bundle-sheath cells of C_4 plants, carries out a reaction that has a counterpart in the citric acid cycle. What is the analogous reaction? Explain.

Answer Malic enzyme catalyzes an oxidative decarboxylation of a hydroxycarboxylic acid in the C_4 pathway:

$$^-OOC-CH(OH)-CH_2-COO^- + NADP^+ \longrightarrow {}^-OOC-CO-CH_3 + CO_2 + NADPH + H^+$$
$$\text{Malate} \qquad\qquad\qquad\qquad\qquad\qquad \text{Pyruvate}$$

which is analogous to the reaction catalyzed in the citric acid cycle by the enzyme isocitrate dehydrogenase:

$$\text{Isocitrate} + NAD^+ \longrightarrow \alpha\text{-ketoglutarate} + NADH + H^+$$

17. *Sucrose and Dental Caries* The most prevalent infection in humans worldwide is dental caries, which stems from the colonization and destruction of tooth enamel by a variety of acidifying microorganisms. These organisms synthesize and live within a water-insoluble network of dextrans, called dental plaque, composed of ($\alpha 1 \rightarrow 6$)-linked polymers of glucose with many ($\alpha 1 \rightarrow 3$) branch points. Polymerization of dextran requires dietary sucrose, and the reaction is catalyzed by a bacterial enzyme, dextran-sucrose glucosyltranferase.

(a) Write the overall reaction for dextran polymerization.

(b) In addition to providing a substrate for the formation of dental plaque, how does dietary sucrose also provide oral bacteria with an abundant source of metabolic energy?

Answer

(a)   Sucrose + $(glucose)_n$ $\longrightarrow$ $(glucose)_{n+1}$ + fructose
                      $dextran_n$ $dextran_{n+1}$

(b)   The fructose generated in the synthesis of dextran is readily taken up by the bacteria and metabolized to acidic compounds.

18. *Regulation of Carbohydrate Synthesis in Plants* Sucrose synthesis occurs in the cytosol, and starch synthesis occurs in the chloroplast stroma; yet both reactions are intricately balanced.

(a) What factors shift the reactions in favor of starch synthesis?

(b) What factors shift the reactions to favor sucrose synthesis?

(c) Given that these two synthetic pathways occur in separate cellular compartments, what enables the two processes to influence each other?

Answer

(a)   Low levels of P_i in the cytosol and high levels of triose phosphate in the chloroplast favor formation of starch.

(b)   High levels of triose phosphate in the cytosol favor formation of sucrose.

(c)   The P_i-triose phosphate antiport system.

CHAPTER **20** **Lipid Biosynthesis**

1. *Pathway of Carbon in Fatty Acid Synthesis* Using your knowledge of fatty acid biosynthesis, provide an explanation for the following experimental observations:

   (a) The addition of uniformly labeled [^{14}C]acetyl-CoA to a soluble liver fraction yields palmitate uniformly labeled with ^{14}C.

   (b) However, the addition of a *trace* of uniformly labeled [^{14}C]acetyl-CoA in the presence of an excess of unlabeled malonyl-CoA to a soluble liver fraction yields palmitate labeled with ^{14}C only in C-15 and C-16.

   Answer Recall that the "loading" of the acyl carrier protein requires an initial addition of acetyl-CoA followed by the addition of malonyl-CoA. Malonyl-CoA is normally produced by the addition of CO_2 to acetyl-CoA by acetyl-CoA carboxylase. Consequently, in the presence of an excess of [^{14}C]acetyl-CoA (**a**), the metabolic pool of malonyl-CoA becomes labeled at C-1 and C-2. This results in the formation of uniformly labeled palmitate. If a *trace* of [^{14}C]acetyl-CoA in the presence of a large excess of unlabeled malonyl-CoA is introduced (**b**), however, the metabolic pool of malonyl-CoA does not become labeled because the trace of [^{14}C]acetyl-CoA is loaded onto the acyl carrier protein (to become C-15 and C-16 of palmitate) rather than transformed into malonyl-CoA (a slow, rate-controlling process). In addition, any labeled malonyl-CoA is diluted by the presence of excess unlabeled malonyl-CoA.

2. *Synthesis of Fatty Acids from Glucose* After a person has consumed large amounts of sucrose, the glucose and fructose that exceed caloric requirements are transformed to fatty acids for triacylglycerol synthesis. This fatty acid synthesis consumes acetyl-CoA, ATP, and NADPH. How are these substances produced from glucose?

   Answer Glucose and fructose are degraded to pyruvate via glycolysis in the cytosol. The pyruvate enters the mitochondrion and is oxidatively decarboxylated to acetyl-CoA, some of which enters the citric acid cycle. The reducing equivalents (NADH, FADH$_2$) produced in the citric acid cycle are used for ATP production by oxidative phosphorylation. The remainder of the acetyl-CoA is exported to the cytosol via the acetyl group shuttle for fatty acid synthesis (see Fig. 20-11). Some NADPH is produced in the pentose phosphate pathway (Fig. 14-22) in the cytosol using glucose as a substrate, and some is produced by the action of malic enzyme on cytoplasmic malate.

3. **Net Equation of Fatty Acid Synthesis** Write the net equation for the biosynthesis of palmitate in rat liver, starting from mitochondrial acetyl-CoA and cytosolic NADPH, ATP, and CO_2.

   Answer The majority of acetyl-CoA used in fatty acid synthesis is formed from the oxidation of pyruvate in the mitochondria. Since the mitochondrial membrane is impermeable to acetyl-CoA, transfer across the membrane occurs via the acetyl group shuttle (see Fig. 20-11). This process requires the input of 1 ATP per acetyl-CoA

   $$\text{Acetyl-CoA}_{(mit)} + \text{ATP} + H_2O \longrightarrow \text{acetyl-CoA}_{(cyt)} + \text{ADP} + P_i + H^+$$
   or
   $$8 \text{ Acetyl-CoA}_{(mit)} + 8\text{ATP} + 8H_2O \longrightarrow 8 \text{ acetyl-CoA}_{(cyt)} + 8\text{ADP} + 8P_i + 8H^+$$

   In the synthesis of palmitate, cytosolic acetyl-CoA ("the primer") is condensed with malonyl-CoA, reduced, hydrated, and reduced again. This process is repeated seven times, each time adding an additional acetyl-CoA to the fatty acyl-CoA. The equation for this process is

   $$8 \text{ Acetyl-S-CoA} + 14\text{NADPH} + 6H^+ + 7\text{ATP} + H_2O \longrightarrow$$
   $$\text{palmitate} + 8\text{CoA} + 14\text{NADP}^+ + 7\text{ADP} + 7P_i$$

   (Note that this differs from Eqn 20-3 in the text because the 7 H_2O for hydrolysis of 7 ATP were omitted from Eqn 20-1, as were 7 H^+ in the products of that reaction.)

   The net equation for the overall process in the cytosol:

   $$8 \text{ Acetyl-S-CoA}_{(mit)} + 15\text{ATP} + 14\text{NADPH} + 9H_2O \longrightarrow$$
   $$\text{palmitate} + 8\text{CoA} + 15\text{ADP} + 15P_i + 14\text{NADP}^+ + 2H^+$$

   This may also be viewed as the sum of the following two reactions:

   $$8 \text{ Acetyl-S-CoA} + 14\text{NADPH} + 13H^+ \longrightarrow$$
   $$\text{CH}_3(\text{CH}_2)_{14}\text{COO}^- + 8\text{CoA-SH} + 14\text{NADP}^+ + 6H_2O$$
   and
   $$15\text{ATP} + 15H_2O \longrightarrow 15\text{ADP} + 15P_i + 15H^+$$

4. **Pathway of Hydrogen in Fatty Acid Synthesis** Consider a preparation that contains all the enzymes and cofactors necessary for fatty acid biosynthesis from added acetyl-CoA and malonyl-CoA.

   (a) If [2-^{2}H]acetyl-CoA (labeled with deuterium, the heavy isotope of hydrogen) and an excess of unlabeled malonyl-CoA are added as substrates, how many deuterium atoms are incorporated into every molecule of palmitate? What are their locations? Explain.

(b) If unlabeled acetyl-CoA and [2-²H]malonyl-CoA are added as substrates, how many deuterium atoms are incorporated into every molecule of palmitate? What are their locations? Explain.

$$^-OOC-\underset{\underset{\textstyle ^2H}{|}}{\overset{\overset{\textstyle ^2H}{|}}{C}}-C\underset{\textstyle S\text{-}CoA}{\overset{\textstyle O}{\diagup\!\!\!\diagdown}}$$

Answer

(a) Three deuteriums per palmitate; all located on C-16; all other two-carbon units are derived from unlabeled malonyl-CoA. (Note that the unlabeled malonyl-CoA is in excess.)

(b) Seven deuteriums per palmitate; one on each even-numbered carbon except C-16. One deuterium is lost from each labeled even-numbered carbon at the dehydration step (see Fig. 20-5).

5. *Energetics of β-Ketoacyl-ACP Synthase* In the condensation reaction catalyzed by β-ketoacyl-ACP synthase (Fig. 20-5), a four-carbon unit is synthesized by the combination of a two-carbon unit and a three-carbon unit, with the release of CO_2. What is the thermodynamic advantage of this process over one that simply combines two two-carbon units?

> *Answer* If a four-carbon unit were synthesized by combining two two-carbon units, the reaction would be reversible—for example, the reaction catalyzed by thiolase in the oxidation of fatty acids (see Fig. 16-8).
>
> $CoA\text{-}SH + CH_3\text{-}CO\text{-}CH_2\text{-}CO\text{-}S\text{-}CoA \rightleftharpoons CH_3\text{-}CO\text{-}S\text{-}CoA + CH_3\text{-}CO\text{-}S\text{-}CoA$
>
> By using the three-carbon unit malonyl-CoA, the activated form of acetyl-CoA (recall that malonyl-CoA synthesis requires the input of energy from ATP), metabolite flow is driven in the direction of fatty acid synthesis by the exergonic release of CO_2.

6. *Modulation of Acetyl-CoA Carboxylase* Acetyl-CoA carboxylase is the principal regulation point in the biosynthesis of fatty acids. Some of the properties of the enzyme are described below:

(a) The addition of citrate or isocitrate raises the V_{max} of the enzyme by as much as a factor of 10.

(b) The enzyme exists in two interconvertible forms that differ markedly in their activities:

> Protomer (inactive) $\rightleftharpoons$ filamentous polymer (active)

Citrate and isocitrate bind preferentially to the filamentous form, and palmitoyl-CoA binds preferentially to the protomer.

Explain how these properties are consistent with the regulatory role of acetyl-CoA carboxylase in the biosynthesis of fatty acids.

Answer The rate-limiting step in the biosynthesis of fatty acids is the carboxylation of acetyl-CoA catalyzed by acetyl-CoA carboxylase. High levels of citrate and isocitrate indicate that conditions are favorable for fatty acid synthesis: an active citric acid cycle is providing a plentiful supply of ATP, reducing equivalents, and acetyl-CoA. Citrate or isocitrate stimulate (increase the V_{max} of) the rate-limiting enzymatic step in fatty acid biosynthesis (**a**). Furthermore, since citrate and isocitrate bind more tightly to the filamentous form of the enzyme (the active form), the presence of citrate or isocitrate drives the protomer $\rightleftharpoons$ polymer equilibrium in the direction of the active (polymer) form (**b**). In contrast, palmitoly-CoA (the end product of fatty acid biosynthesis) drives the equilibrium in the direction of the inactive (protomer) form. Hence, when the end product of fatty acid biosynthesis builds up, fatty acid biosynthesis is slowed down.

7. ***Shuttling of Acetyl Groups across the Inner Mitochondrial Membrane*** The acetyl group of acetyl-CoA, produced by the oxidative decarboxylation of pyruvate in the mitochondrion, is transferred to the cytosol by the acetyl group shuttle outlined in Figure 20-11.

(a) Write the overall equation for the transfer of one acetyl group from the mitochondrion to the cytosol.

(b) What is the cost of this process in ATPs per acetyl group?

(c) In Chapter 16 we encountered an acyl group shuttle in the transfer of fatty acyl-CoA from the cytosol to the mitochondrion in preparation for β oxidation (see Fig. 16-6). One result of that shuttle was separation of the mitochondrial and cytosolic pools of CoA. Does the acetyl group shuttle also accomplish this?

Answer
(a) The reactions involved in the transfer of one acetyl group:

Acetyl-CoA$_{(mit)}$ + oxaloacetate$_{(mit)}$ $\longrightarrow$ citrate$_{(mit)}$ + CoA$_{(mit)}$

Citrate$_{(mit)}$ $\longrightarrow$ citrate$_{(cyt)}$

Citrate$_{(cyt)}$ + ATP + CoA$_{(cyt)}$ $\longrightarrow$ oxaloacetate$_{(cyt)}$ + ADP + P$_i$ + acetyl-CoA$_{(cyt)}$

Oxaloacetate$_{(cyt)}$ + NADH + H$^+$ $\longrightarrow$ malate$_{(cyt)}$ + NAD$^+$

Malate$_{(cyt)}$ $\longrightarrow$ malate$_{(mit)}$

Malate$_{(mit)}$ + NAD$^+$ $\longrightarrow$ oxaloacetate$_{(mit)}$ + NADH + H$^+$

The overall equation:

Acetyl-CoA$_{(mit)}$ + ATP + CoA$_{(cyt)}$ $\longrightarrow$ acetyl-CoA$_{(cyt)}$ + ADP + P$_i$ + CoA$_{(mit)}$

(b) The transfer of one acetyl group from the mitochondrial matrix to the cytosol (one turn of the acetyl group shuttle) is accompanied by the hydrolysis of ATP.

(c) The transfer requires the input of one cytosolic CoA, and one mitochondrial CoA is released. Thus, the shuttle keeps the cytosolic and mitochondrial pools of CoA separate.

8. *Oxygen Requirement for Desaturases* The biosynthesis of palmitoleate (Fig. 20-14), a common unsaturated fatty acid with a cis double bond in the Δ^9 position, uses palmitate as a precursor. Can this be carried out under strictly anaerobic conditions? Explain.

> *Answer* No; oxygen is required. The double bond in palmitoleic acid is introduced by an oxidation catalyzed by fatty acyl-CoA oxygenase (or fatty acyl-CoA desaturase; see Fig. 20-14), a mixed-function oxidase that requires molecular oxygen as a cosubstrate.

9. *Energy Cost of Triacylglycerol Synthesis* Use a net equation for the biosynthesis of tripalmitoylglycerol (tripalmitin) from glycerol and palmitate to show how many ATPs are required per molecule of tripalmitin formed.

> *Answer* The reactions involved in conversion of glycerol and palmitate to tripalmitin:
>
> ATP^{4-} + glycerol $\longrightarrow$ glycerol-3-phosphate^{2-} + ADP^{3-} + H^+
>
> 3 Palmitate + $3ATP^{4-}$ + 3CoA $\longrightarrow$ 3 palmitoyl-CoA + $3AMP^{2-}$ + $3PP_i^{3-}$ + $3H^+$
>
> $3H_2O$ + $3PP_i^{3-}$ $\longrightarrow$ $6P_i^{2-}$ + $3H^+$
>
> 2 Palmitoyl-CoA + glycerol-3-phosphate^{2-} $\longrightarrow$ dipalmitoylglycerol-3-phosphate^{2-} + 2CoA
>
> Dipalmitoyl-3-phosphate^{2-} + H_2O $\longrightarrow$ dipalmitoylglycerol + P_i^{2-}
>
> Dipalmitoylglycerol + palmitoyl-CoA $\longrightarrow$ tripalmitin + CoA
>
> The overall reaction:
>
> 3 Palmitate + glycerol + $4ATP^{4-}$ + $4H_2O$ $\longrightarrow$
> $$\text{tripalmitin} + 3AMP^{2-} + ADP^{3-} + 7P_i^{2-} + 7H^+$$
>
> Including
>
> $3AMP^{2-}$ + $3ATP^{4-}$ $\longrightarrow$ $6ADP^{3-}$
>
> The overall reaction becomes
>
> 3 Palmitate + glycerol + $7ATP^{4-}$ + $4H_2O$ $\longrightarrow$ tripalmitin + $7ADP^{3-}$ + $7P_i^{2-}$ + $7H^+$

10. *Turnover of Triacylglycerols in Adipose Tissue* When [^{14}C]glucose is added to the balanced diet of adult rats, there is no increase in the total amount of stored triacylglycerols, but the triacylglycerols become labeled with ^{14}C. Explain.

> *Answer* In adult rats, stored triacylglycerols are maintained at a steady-state level through a balance of the rates of degradation and biosynthesis. Hence, the depots of fats are continually turned over, explaining the incorporation of ^{14}C from dietary [^{14}C] glucose.

11. ***Energy Cost of Phosphatidylcholine Synthesis***   Write the sequence of steps and the net reaction for the biosynthesis of phosphatidylcholine by the salvage pathway from oleate, palmitate, dihydroxyacetone phosphate, and choline. Starting from these precursors, what is the cost in number of ATPs of the synthesis of phosphatidylcholine by the salvage pathway?

> ***Answer***   The sequence of steps required to carry out phosphatidylcholine biosynthesis:
>
> Dihydroxyacetone phosphate^{2-} + NADH + H$^+$ $\longrightarrow$ glycerol-3-phosphate + NAD$^+$
>
> Palmitate + ATP + CoA $\longrightarrow$ palmitoyl-CoA + AMP + PP$_i$
>
> Oleic acid + ATP + CoA-SH $\longrightarrow$ oleoyl-CoA + AMP + PP$_i$
>
> 2PP$_i$ + 2H$_2$O $\longrightarrow$ 2P$_i$
>
> Glycerol-3-phosphate + palmitoyl-CoA + oleoyl-CoA $\longrightarrow$ L-phosphatidate + 2CoA
>
> L-Phosphatidate + H$_2$O $\longrightarrow$ 1,2-diacylglycerol + P$_i$
>
> ATP + choline $\longrightarrow$ ADP + phosphocholine
>
> CTP + phosphocholine $\longrightarrow$ CDP-choline + PP$_i$
>
> PP$_i$ + H$_2$O $\longrightarrow$ 2P$_i$
>
> CDP-choline + 1,2-diacylglycerol $\longrightarrow$ CMP + phosphatidylcholine
>
> The net reaction:
>
> Dihydroxyacetone phosphate^{2-} + NADH + palmitate + oleic acid + 3ATP^{4-} +
>   CTP^{4-} + choline$^+$ + 4H$_2$O $\longrightarrow$
>   phosphatidylcholine + NAD$^+$ + 2AMP^{2-} + ADP^{3-} + CMP^{2-} + 5P$_i^{2-}$ + H$^+$
>
> This is can also be viewed as the sum of the following two equations:
>
> Dihydroxyacetone phosphate^{2-} + palmitate + oleic acid + choline$^+$ + NADH + 2H$^+$ $\longrightarrow$
>   phosphatidylcholine + NAD$^+$ + 3H$_2$O
>
> and
>
> 3ATP^{4-} + CTP^{4-} + 7H$_2$O $\longrightarrow$ 2AMP^{2-} + ADP^{3-} + CMP^{2-} + 5P$_i^{2-}$ + 3H$^+$
>
> Note that the net equation shows the production of three nucleoside monophosphates (2 AMP + CMP). To recycle these monophosphates to nucleoside diphosphates requires the input of 3 ATP.
>
>   2AMP + 2ATP $\rightleftharpoons$ 4ADP
>
>   CMP + ATP $\rightleftharpoons$ CDP + ADP
>
> Furthermore, if we consider that CTP is energetically equivalent to ATP, then the synthesis of one phosphatidylcholine requires a total of seven ATP.

12. ***Salvage Pathway for Synthesis of Phosphatidylcholine***   A young rat maintained on a diet deficient in methionine fails to thrive unless choline is included in the diet. Explain.

> ***Answer***   The rat has two pathways for synthesizing phosphatidylcholine: the de novo pathway and the salvage pathway. The de novo pathway requires the transfer of a methyl group from *S*-adenosylmethionine (adoMet; see Fig. 20-25) to phosphatidylethanolamine. When the diet is deficient in methionine (an essential amino acid), the biosynthesis of adoMet and phosphatidylcholine is severely impaired. The salvage pathway, on the other hand, does not require adoMet but utilizes available choline. Thus, phosphatidylcholine can be synthesized when the diet is deficient in methionine as long as choline is available.

13. **Synthesis of Isopentenyl Pyrophosphate** If 2-[^{14}C]acetyl-CoA is added to a rat liver homogenate that is synthesizing cholesterol, where will the ^{14}C label appear in Δ^3-isopentenyl pyrophosphate, the activated form of an isoprene unit?

Answer Three acetyl units are required for the synthesis of an isoprene unit (see Fig. 20-31). The ^{14}C label at C-2 of acetyl-CoA ends up in three places in the activated isoprene unit:

$$^{14}CH_2 \atop {\diagdown \atop C - ^{14}CH_2 - CH_2 - \atop \diagup} \atop ^{14}CH_3$$

14. **HMG-CoA in Ketone Body Synthesis** The rate-limiting step in the early stages of cholesterol biosynthesis is the conversion of β-hydroxy-β-methyglutaryl-CoA to mevalonate, catalyzed by HMG-CoA reductase (Fig. 20-32). The liver of a fasting animal has decreased reductase activity. When the flow through this reaction is reduced, what is the effect on the formation of ketone bodies from acetyl-CoA? How does this explain increased ketosis during fasting? (Hint: See Figure 16-16.)

Answer The pathways to mevalonate and ketone bodies (acetoacetate, acetone, and D-β-hydroxybutyrate) have many steps in common. A decrease in the activity of HMG-CoA reductase results in an accumulation of β-hydroxy-β-methylglutaryl-CoA. Since this is a substrate for hydroxymethylglutaryl-CoA lyase (see Fig. 16-16), decreased reductase activity leads to the increased production of acetoacetate—that is, ketosis.

15. **Activated Donors in Lipid Synthesis** In the biosynthesis of complex lipids, components are assembled by transfer of the appropriate group from an activated donor. For example, the activated donor of acetyl groups is acetyl-CoA. For each of the following groups, give the form of the activated donor: (a) phosphate; (b) D-glucosyl; (c) phosphoethanolamine; (d) D-galactosyl; (e) fatty acyl; (f) methyl; (g) the two-carbon group in fatty acid biosynthesis; (h) Δ^3-isopentenyl.

Answer (a) ATP; (b) UDP-glucose; (c) CDP-ethanolamine; (d) UDP-galactose; (e) fatty acyl-CoA; (f) S-adenosylmethionine; (g) malonyl-CoA; (h) Δ^3-isopentenyl pyrophosphate.

CHAPTER 21 **Biosynthesis of Amino Acids, Nucleotides, and Related Molecules**

1. *Cofactors for One-Carbon Transfer Reactions* Most one-carbon transfers are promoted by one of three cofactors: biotin, tetrahydrofolate, or *S*-adenosylmethionine (Chapter 17). *S*-Adenosylmethionine is used as a methyl group donor in most reactions; the transfer potential of the methyl group in N^5-methyltetrahydrofolate is insufficient for most biosynthetic reactions. However, one example of the use of N^5-methyltetrahydrofolate in a methyl group transfer occurs in the methionine synthase reaction (step 9 of Fig. 21-12), and methionine is the immediate precursor of *S*-adenosylmethionine (see Fig. 17-20). Explain how the methyl group of *S*-adenosylmethionine can be derived from N^5-methyltetrahydrofolate, even though the transfer potential of the methyl group in N^5-methyltetrahydrofolate is 10^3 times *lower* than that in *S*-adenosylmethionine.

> *Answer* The transfer potential of the methyl group of N^5-methyltetrahydrofolate is quite sufficient for the synthesis of methionine, which has an even lower methyl group transfer potential. The methyl group is activated by addition of the adenosyl group from ATP when methionine is converted to *S*-adenosylmethionine (see Fig. 17-20). Recall that *S*-adenosylmethionine synthesis is one of only two known biochemical reactions in which triphosphate is released from ATP. Hydrolysis of the triphosphate renders the reaction thermodynamically more favorable.

2. *Defect in Phenylalanine Hydroxylase and Diet* Tyrosine is normally a nonessential amino acid, but individuals with a genetic defect in phenylalanine hydroxylase require tyrosine in their diet for normal growth. Explain.

> *Answer* In animals, tyrosine is synthesized from phenylalanine by phenylalanine hydroxylase. If this enzyme is defective, the biosynthetic route to Tyr is blocked and Tyr must be obtained from the diet.

3. *Equation for the Synthesis of Aspartate from Glucose*   Write the net equation for the synthesis of the nonessential amino acid aspartate from glucose, carbon dioxide, and ammonia.

> *Answer*   We can approach this problem by working "backwards" from aspartate to glucose as follows. Aspartate is synthesized from oxaloacetate by transamination from glutamate; the glutamate is synthesized from α-ketoglutarate by glutamate dehydrogenase:
>
> Oxaloacetate + glutamate $\longrightarrow$ aspartate + α-ketoglutarate
>
> α-Ketoglutarate + NH_3 + $2H^+$ + NADH $\longrightarrow$ glutamate + NAD^+ + H_2O
>
> The sum of these reactions is
>
> Oxaloacetate + NH_3 + $2H^+$ + NADH $\longrightarrow$ aspartate + NAD^+ + H_2O
>
> Recall from Chapter 15 that oxaloacetate is synthesized from pyruvate by pyruvate carboxylase, and from Chapter 14 that pyruvate is synthesized from glucose via glycolysis:
>
> Pyruvate + CO_2 + ATP + H_2O $\longrightarrow$ oxaloacetate + ADP + P_i + $2H^+$
>
> Glucose + $2NAD^+$ + 2ADP + $2P_i$ $\longrightarrow$ 2 pyruvate + 2NADH + $2H^+$ + 2ATP + $2H_2O$
>
> Thus we can write the net equation for aspartate synthesis:
>
> Glucose + $2CO_2$ + $2NH_3$ $\longrightarrow$ 2 aspartate + $2H^+$ + $2H_2O$

4. *Inhibition of Nucleotide Synthesis by Azaserine*   The diazo compound *O*-(2-diazoacetyl)-L-serine, known also as azaserine (Fig. 21-41), is a powerful inhibitor of those enzymes that transfer ammonia from glutamine to an acceptor (amidotransferases) during biosynthesis. If growing cells are treated with azaserine, what intermediates in nucleotide biosynthesis would you expect to accumulate? Explain.

> *Answer*   In the de novo pathway of purine biosynthesis, the first step that requires glutamine is the conversion of 5-phosphoribosyl-1-pyrophosphate (PRPP) to 5-phospho-β-D-ribosylamine. In the presence of azaserine, which inhibits this conversion, PRPP will accumulate.

5. *Nucleotide Biosynthesis in Amino Acid Auxotrophic Bacteria*   Although normal *E. coli* cells can synthesize all the amino acids, some mutants, called amino acid auxotrophs, are unable to synthesize specific amino acids and require the addition of that amino acid to the culture medium for optimal growth. In addition to their role in protein synthesis, specific amino acids are also required in the biosynthesis of other nitrogenous cell products. Consider the three amino acid auxotrophs that are unable to synthesize glycine, glutamine, and aspartate, respectively. For each mutant what nitrogenous cell products other than proteins would fail to be synthesized?

> *Answer*   Glycine, glutamine, and aspartate are required for the de novo synthesis of purine nucleotides; aspartate is required for the de novo synthesis of UMP; aspartate and glutamine are required for the de novo synthesis of CTP. Thus, glycine auxotrophs would fail to synthesize adenine and guanine nucleotides. Glutamine auxotrophs would fail to synthesize adenine, guanine, and cytosine nucleotides. Aspartate auxotrophs would fail to synthesize adenine, guanine, cytosine, and uridine nucleotides.

6. *Inhibitors of Nucleotide Biosynthesis*   Suggest mechanisms for the inhibition of **(a)** alanine racemase by L-fluoroalanine and **(b)** glutamine amidotransferases by azaserine.

> **Answer**
> **(a)**   See Figure 17-7 (step 2) for the reaction mechanism of amino acid racemization. The F atom of fluoroalanine is an excellent leaving group. Fluoroalanine causes irreversible (covalent) inhibition of alanine racemase. One plausible mechanism is:

> **(b)**   Azaserine (see Fig. 21-41) is an analog of glutamine. The diazoacetyl group is highly reactive and forms covalent bonds with nucleophiles at the active site of glutamine amidotransferases.

7. *Nucleotides Are Poor Sources of Energy*   In most organisms, nucleotides are not employed as energy-yielding fuels. What observations support this conclusion? Why are nucleotides relatively poor sources of energy in mammals?

> **Answer** Organisms do not store nucleotides to be used as fuel and do not completely degrade them, but rather hydrolyze them to release the bases, which can be recovered in salvage pathways. The low C:N ratio of nucleotides makes them poor sources of energy.

8. *Mode of Action of Sulfa Drugs*   Some bacteria require the inclusion of *p*-aminobenzoate in the culture medium for normal growth. The growth of such bacteria is severely inhibited by the addition of sulfanilamide, one of the earliest antibacterial sulfa drugs. Moreover, in its presence, 5′-aminoimidazole-4-carboxamide ribonucleotide (AICAR; see Fig. 21-27) accumulates in the culture medium. Both effects are reversed by the addition of excess *p*-aminobenzoate.

p-Aminobenzoate   Sulfanilamide

(a) What is the role of *p*-aminobenzoate? (Hint: See Fig. 17-18).

(b) Why does AICAR accumulate in the presence of sulfanilamide?

(c) Why is the inhibition and accumulation reversed by the addition of excess *p*-aminobenzoate?

Answer

(a) *p*-Aminobenzoate is a component of tetrahydrofolate (see Fig. 17-18) and its derivative, N^5N^{10}-methylenetetrahydrofolate, the cofactor involved in the transfer of one-carbon units.

(b) Sulfanilamide is a structural analog of *p*-aminobenzoate. In the presence of sulfanilamide, bacteria are unable to synthesize tetrahydrofolate, a cofactor necessary for the transformation of AICAR to *N*-formylaminoimidazole-4-carboxamide ribonucleotide (FAICAR) by the addition of —CHO; AICAR accumulates.

(c) Excess *p*-aminobenzoate reverses the growth inhibition and the ribunucleotide accumulation by competing with sulfanilamide for the active site of the enzyme involved in tetrahydrofolate biosynthesis. The competitive inhibition by sulfanilamide is overcome by the addition of excess substrate (*p*-aminobenzoate).

9. **Treatment of Gout** Allopurinol (Fig. 21-40), an inhibitor of xanthine oxidase, is used to treat chronic gout. Explain the biochemical basis for this treatment. Patients treated with allopurinol sometimes develop xanthine stones in the kidneys, although the incidence of kidney damage is much lower than in untreated gout. Explain this observation in light of the following solubilities in urine: uric acid, 0.15 g/L; xanthine, 0.05 g/L; and hypoxanthine, 1.4 g/L.

Answer Treatment with allopurinol has two biochemical consequences. (1) Conversion of hypoxanthine to uric acid is inhibited, causing accumulation of hypoxanthine, which is more soluble than uric acid and more readily excreted. This alleviates the clinical problems associated with AMP degradation. (2) Conversion of guanine to uric acid is also inhibited, causing accumulation of xanthine, which is, unfortunately, even less soluble than uric acid. This is the source of xanthine stones. Because less GMP than AMP is degraded, kidney damage caused by xanthine stones is less than that caused by untreated gout.

10. **ATP Consumption by Root Nodules in Legumes** The bacteria residing in the root nodules of the pea plant consume more than 20% of all the ATP produced by the plant. Suggest a reason why these bacteria consume so much ATP.

Answer The bacteria in the root nodules maintain a symbiotic relationship with the plant: the plant supplies ATP and reducing power, and the bacteria supply ammonium ion by reducing atmospheric nitrogen. This reduction requires large quantities of ATP.

11. *Pathway of Carbon in Pyrimidine Biosynthesis* What are the locations of ^{14}C in the orotate molecule present in cells grown on a small amount of uniformly labeled [^{14}C]succinate? Explain.

    Answer The ^{14}C-labeled orotate arises from the following pathway:

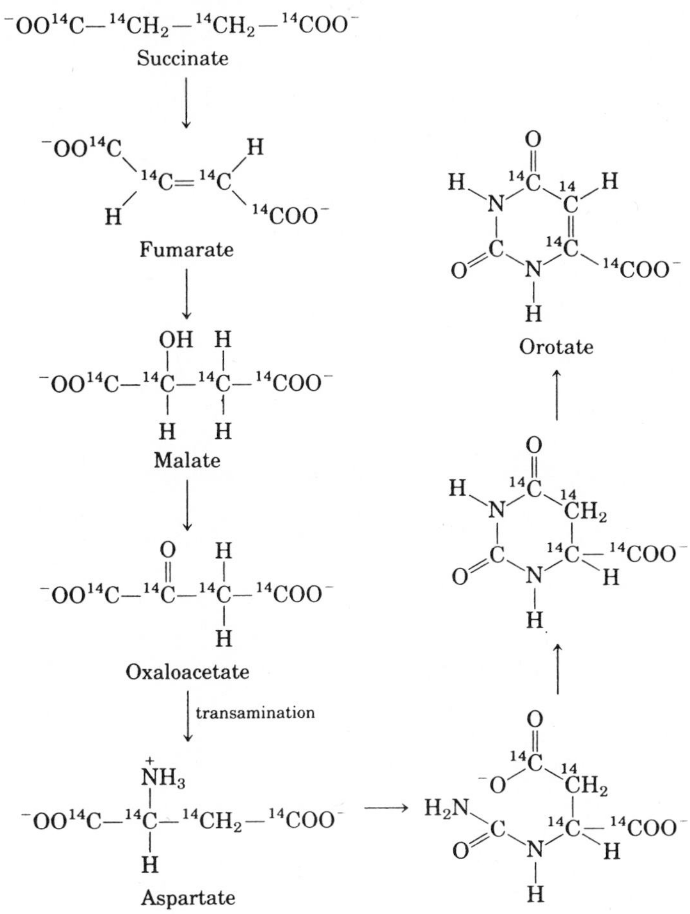

$^-OO^{14}C-^{14}CH_2-^{14}CH_2-^{14}COO^-$

Succinate

Fumarate

Malate

Oxaloacetate

transamination

Aspartate

Orotate

CHAPTER **22**   **Integration and Hormonal Regulation of Mammalian Metabolism**

1. ***ATP and Phosphocreatine as Sources of Energy for Muscle*** In contracting skeletal muscle, the concentration of phosphocreatine drops while the concentration of ATP remains fairly constant. Explain how this happens.

In a classic experiment, Robert Davies found that if the muscle is first treated with 1-fluoro-2,4-dinitrobenzene (see Fig. 5-14), the concentration of ATP in the muscle declines rapidly, whereas the concentration of phosphocreatine remains unchanged during a series of contractions. Suggest an explanation.

    Answer Muscle contraction leads to the net hydrolysis of ATP. Although the amount of ATP in muscle is very small, the supply can be rapidly replenished by phosphate transfer from the phosphocreatine reservoir, catalyzed by creatine kinase

        Phosphocreatine + ADP $\rightleftharpoons$ creatine + ATP

Since this reaction is rapid relative to ATP utilization by muscle, ATP concentration remains in a steady state. The effects of pretreatment with fluoro-2,4-dinitrobenzene suggest that this is an effective inhibitor of creatine kinase. Under working conditions, the small amount of muscle ATP is quickly depleted and cannot be replenished.

2. ***Metabolism of Glutamate in the Brain*** Glutamate in the blood flowing into the brain is transformed into glutamine, which appears in the blood leaving the brain. What is accomplished by this metabolic conversion? How does it take place? Actually, the brain can generate more glutamine than can be made from the glutamate entering in the blood. How does this extra glutamine arise? (Hint: You may want to review amino acid catabolism in Chapter 17. Recall that NH_3 is very toxic to the brain.)

    Answer Glutamate is produced from transamination of α-ketoglutarate that arises from the breakdown of glucose via glycolysis and the citric acid cycle. Ammonia is removed by conversion of glutamate to glutamine. The glutamate thus produced reacts with a second ammonia to yield glutamine. Glutamine is transported to the liver and converted into urea.

3. *Absence of Glycerol Kinase in Adipose Tissue* Glycerol-3-phosphate is a key intermediate in the biosynthesis of triacylglycerols. Adipocytes, which are specialized for the synthesis and degradation of triacylglycerols, cannot directly use glycerol because they lack glycerol kinase, which catalyzes the reaction

$$\text{Glycerol} + \text{ATP} \longrightarrow \text{glycerol-3-phosphate} + \text{ADP}$$

How does adipose tissue obtain the glycerol-3-phosphate necessary for triacylglycerol synthesis? Explain.

> *Answer* The glycerol-3-phosphate arises from glucose through the glycolytic pathway (Chapter 14). The glycolytic intermediate dihydroxyacetone phosphate is reduced to glycerol-3-phosphate by the NADH-requiring enzyme glycerol phosphate dehydrogenase.

4. *Hyperglycemia in Patients with Acute Pancreatitis* Patients with acute pancreatitis are treated by withholding protein from the diet and by intravenous administration of glucose-saline solution. What is the biochemical basis for these measures? Patients undergoing this treatment commonly experience hyperglycemia. Why?

> *Answer* Acute pancreatitis is caused by the premature activation of the proenzymes secreted by the pancreas, including proteases, which results in the self-degradation (autolysis) of pancreatic tissue. The amount of pancreatic secretion is governed by the type and amount of food ingested and can be minimized if the patient excludes protein from the diet. Any reduction of fluid, caloric, and electrolyte intake can be compensated by the intravenous administration of glucose-saline solution. Hyperglycemia apparently results from the decreased ability of irritated pancreatic tissue to release insulin, leading to reduced glucose utilization.

5. *Oxygen Consumption during Exercise* A sedentary adult consumes about 0.05 L of O_2 during a 10 s period. A sprinter, running a 100 m race, consumes about 1 L of O_2 during the same time period. After finishing the race, the sprinter will continue to breathe at an elevated but declining rate for some minutes, consuming an extra 4 L of O_2 above the amount consumed by the sedentary individual.

(a) Why do the O_2 needs increase dramatically during the sprint?

(b) Why do the O_2 demands remain high after the sprint is completed?

> *Answer* Resting muscle requires a minimal flux of ATP, which is provided by the oxidation (measured by O_2 consumption) of free fatty acids and ketone bodies from the liver. Increased muscular activity increases the demand for ATP, which is met by increased O_2 consumption.
>
> **(a)** During a sprint, muscle tissue is maximally active, and the rate at which oxygen and fuels are supplied is insufficient to meet the required ATP. Under these conditions, muscle tissue generates ATP from the anaerobic transformation of glycogen to lactate.
>
> **(b)** After completion of the sprint, lactate is transported to the liver, where it is converted back to glucose and glycogen. This process requires ATP and thus requires oxygen above the amount necessary in the resting state.

6.   ***Thiamin Deficiency and Brain Function***   Individuals with thiamin deficiency display a number of characteristic neurological signs: loss of reflexes, anxiety, and mental confusion. Suggest a reason why thiamin deficiency is manifested by changes in brain function.

>   ***Answer***   The brain uses glucose as its primary fuel. Glucose is oxidized via the glycolytic sequence and the citric acid cycle, and a key reaction in this process is the thiamine pyrophosphate-dependent oxidative decarboxylation of pyruvate to acetyl-CoA. Thus, the neurological symptoms associated with thiamine deficiency suggest a reduced glucose utilization by the brain, with associated changes in brain function.

7.   ***Significance of Hormone Concentration***   Under normal conditions, the human adrenal medulla secretes epinephrine ($C_9H_{13}NO_3$) at a rate sufficient to maintain a concentration of 10^{-10} M in the circulating blood. To appreciate what that concentration means, calculate the diameter of a round swimming pool, with a water depth of 2 m, that would be needed to dissolve 1 g (about 1 teaspoon) of epinephrine to a concentration equal to that in blood.

>   ***Answer***   Concentration of epinephrine in blood = (183 g/mol)(10^{-10} mol/L) = 183 x 10^{-10} g/L. Thus 1 g must be dissolved in
>
>   (1 g)/(183 x 10^{-10} g/L) =   5.46 x 10^7 L
>
>   Volume of pool = $\pi r^2 d$ = 3.142(2m)r^2
>   Since 1 L = 10^{-3} m^3, the volume of water required is 5.46 x 10^4 m^3. Thus
>
>   3.142(r^2)(2 m) = 5.46 x 10^4 m^3
> $\qquad\quad r^2$ = 0.869 x 10^4 m^2
> $\qquad\quad r$ = 0.93 x 10^2 m
>
>   The diameter of the pool = $2r$ = 1.86 x 10^2 m ≈ 200 m

8.   ***Regulation of Hormone Levels in the Blood***   The half-life of most hormones in the blood is relatively short. For example, if radioactively labeled insulin is injected into an animal, one can determine that within 30 min half the hormone has disappeared from the blood.

(a)   What is the importance of the relatively rapid inactivation of circulating hormones?

(b)   In view of this rapid inactivation, how can the circulating hormone level be kept constant under normal conditions?

(c)   In what ways can the organism make possible rapid changes in the level of circulating hormones?

>   ***Answer***
>   (a)   The ability to rapidly inactivate circulating hormones serves as an efficient regulatory mechanism: the organism can change hormone concentrations rapidly in accord with sudden changes in environmental conditions.
>
>   (b)   Under normal conditions, hormone concentrations are kept constant by maintaining the rate of hormone synthesis equal to the rate of hormone degradation.
>
>   (c)   Other means to vary hormone concentrations include changes in the rates of release from storage, transport, and conversion from prohormone to active hormone.

9. *Water-Soluble versus Lipid-Soluble Hormones* On the basis of their physical properties, hormones fall into one of two categories: those that are very soluble in water but relatively insoluble in lipids (e.g., epinephrine) and those that are relatively insoluble in water but highly soluble in lipids (e.g., steroid hormones). In their role as regulators of cellular activity, most water-soluble hormones do not penetrate into the interior of their target cells. The lipid-soluble hormones, by contrast, do penetrate into their target cells and ultimately act in the nucleus. What is the correlation between solubility, the location of receptors, and the mode of action of the two classes of hormones?

> *Answer* Because of their low solubility in a lipid environment, water-soluble hormones cannot penetrate plasma membrane. Rather, they bind to a receptor on the surface of the cell. In the case of epinephrine, this receptor is linked via the G_s protein to an enzyme (adenylate cyclase) that catalyzes the formation of a large amount of second messenger (cAMP) inside the cell. In contrast, lipid-soluble hormones can readily penetrate the hydrophobic core of the plasma membrane. Once inside the cell they can act on receptors directly and thus do not need a second messenger.

10. *Hormone Experiments in Cell-Free Systems* In the 1950s, Earl Sutherland and his colleagues carried out pioneering experiments to elucidate the mechanism of action of epinephrine and glucagon. In the light of our current understanding of hormone action as described in this chapter, interpret each of the experiments described below. Identify the components and indicate the significance of the results.

(a) The addition of epinephrine to a homogenate or broken-cell preparation of normal liver resulted in an increase in the activity of glycogen phosphorylase. However, if the homogenate was first centrifuged at a high speed and epinephrine or glucagon was added to the clear supernatant fraction containing phosphorylase, no increase in phosphorylase activity was observed.

(b) When the particulate fraction sedimented from a liver homogenate by centrifugation was separated and treated with epinephrine, a new substance was produced. This substance was isolated and purified. Unlike epinephrine, this substance activated glycogen phosphorylase when added to the clear supernatant fraction of the homogenate.

(c) The substance obtained from the particulate fraction was heat-stable; that is, heat treatment did not prevent its capacity to activate phosphorylase. (Hint: Would this be the case if the substance were a protein?) The substance appeared nearly identical to a compound obtained when pure ATP was treated with barium hydroxide. (Figure 12-6 will be helpful.)

> *Answer*
> (a) Epinephrine added to a homogenate or broken-cell preparation of normal liver will bind to its receptor, which can then activate adenylate cyclase via a G_s protein (the transducer), resulting in the production of cAMP (second messenger). The receptor, G_s protein, and adenylate cyclase are all membrane-bound proteins, thus centrifugation of the preparation sediments the adenylate kinase activity into the particulate fraction. In contrast, phosphorylase and protein kinase are soluble enzymes found in the supernatant, but phosphorylase activity is no longer affected by addition of either hormone.
>
> (b) Treatment of the particulate fraction with epinephrine stimulates the production of cAMP. The observation that cAMP stimulates glycogen phosphorylase (through the activation of protein kinase) in the absence of the particulate fraction (something that epinephrine or glucagon will not do) provides the important clue that cAMP serves as the second messenger, that is, the intermediate messenger between epinephrine and phosphorylation of glycogen phosphorylase.

(c) The effect of heat treatment on a biological substance is one way to test whether the substance is a protein, which generally inactivate when heated. The observation that the substance is heat-stable suggests that it is not a protein but rather a simple organic molecule. David Lipkin discovered that treatment of ATP with barium hydroxide produces a compound that he identified as adenosine-3',5'-monophosphate (cyclic AMP). The fact that this synthetic substance was virtually identical to the heat-stable substance isolated by Sutherland from liver cells confirmed the chemical structure of the second messenger.

11. **Effect of Dibutyryl-cAMP versus cAMP on Intact Cells** The physiological effects of the hormone epinephrine should in principle be mimicked by the addition of cAMP to the target cells. In practice, the addition of cAMP to intact target cells elicits only a minimal physiological response. Why?

When the structurally related derivative dibutyryl-cAMP (shown below) is added to intact cells, the expected physiological responses can readily be seen. Explain the basis for the difference in cellular response to these two substances. Dibutyryl-cAMP is a widely used derivative in studies of cAMP function.

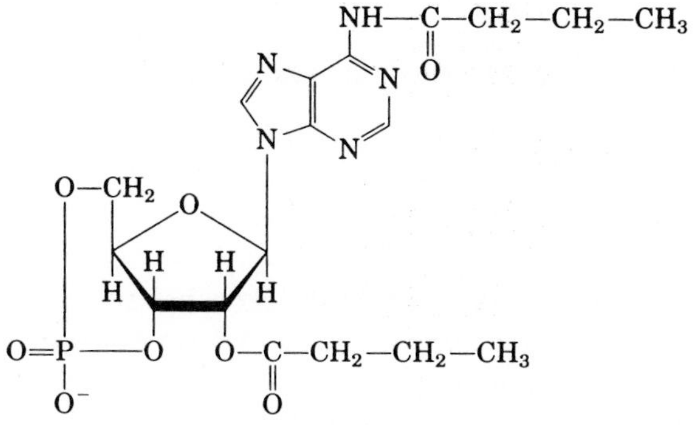

Dibutyryl-cAMP

Answer Unlike epinephrine or glucagon, which bind to their receptors on the outer surfaces of intact cells, cAMP must be inside the cell to carry out its physiological function. Cyclic AMP, a highly polar molecule, is not very soluble in the lipid phase and thus passes through the plasma membrane with difficulty. Dibutyryl-cAMP is much more soluble in the lipid phase because of the presence of two hydrophobic chains, and this markedly enhances its passage into the cell through the plasma membrane.

12. **Effect of Cholera Toxin on Adenylate Cyclase** The gram-negative bacterium *Vibrio cholerae* produces a protein, cholera toxin (M_r 90,000), responsible for the characteristic symptoms of cholera: extensive loss of body water and Na^+ through continuous, debilitating diarrhea. If body fluids and Na^+ are not replaced, severe dehydration will occur; untreated, the disease is often fatal. When the cholera toxin gains access to the human intestinal tract it binds tightly to specific sites in the plasma membrane of the epithelial cells lining the small intestine, causing adenylate cyclase to undergo activation that persists for hours or days.

(a) What is the effect of cholera toxin on the level of cAMP in the intestinal cells?

(b) Based on the information above, can you suggest how cAMP normally functions in intestinal epithelial cells?

(c) Suggest a possible treatment for cholera.

Answer

(a)   Cholera toxin is an enzyme that catalyzes the transfer of ADP-ribose from NAD^+ to the α subunit of the G_s protein, thereby inhibiting the GTPase activity. The G_s α subunit is thus "locked" in its GTP-bound, activating conformation, and consequently adenylate cyclase remains active. This results in abnormally high levels of cAMP in intestinal epithelial cells.

(b)   The clinical observation that the toxin results in movement of Na^+ ions and water from the cells into the gut (diarrhea) suggests that cAMP regulates Na^+ ion permeability in intestinal epithelial cells. As Na^+ ions are expelled into the gut, the change in osmotic potential results in net movement of water into the gut.

(c)   Treatment of the disease consists of continuous replacement of the lost body fluids until the disease runs its course. Eventually the toxin will be excreted. A potentially effective therapy may eventually be derived from biochemical and pharmacological research on inhibitors of the ADP-ribosylating activity of cholera toxin.

13. *Metabolic Differences in Muscle and Liver in a "Fight or Flight" Situation* During a "fight or flight" situation, the release of epinephrine promotes glycogen breakdown in the liver, heart, and skeletal muscle. The end product of glycogen breakdown in the liver is glucose. In contrast, the end product in skeletal muscle is pyruvate.

(a)   Why are different products of glycogen breakdown observed in the two tissues?

(b)   What is the advantage to the organism during a "fight or flight" condition of having these specific glycogen breakdown routes?

Answer

(a)   Heart and skeletal muscle lack the enzyme glucose-6-phosphate phosphatase. Thus any glucose-6-phosphate produced enters the glycolytic pathway to form pyruvate, which under oxygen-deficient conditions is converted to lactate. In the liver, glucose-6-phosphate from glycogen is converted to glucose by the G6P phosphatase.

(b)   The plasma membrane is impermeable to charged species, hence phosphorylated intermediates are retained within the cell. In preparation for a "fight or flight" situation, the concentration of glycolytic intermediates in muscle tissue needs to be high. The liver, on the other hand, supplies the glucose necessary to maintain a steady-state level of blood glucose. In a "fight or flight" situation, glucose needs to be mobilized from the liver to the bloodstream by dephosphorylation, catalyzed by glucose-6-phosphate phosphatase.

14. *Excessive Amounts of Insulin Secretion: Hyperinsulinism* Certain malignant tumors of the pancreas cause excessive production of insulin by the β cells. Affected individuals exhibit shaking and trembling, weakness and fatigue, sweating, and hunger. If this condition is prolonged, brain damage occurs.

(a)   What is the effect of hyperinsulinism on the metabolism of carbohydrate, amino acids, and lipids by the liver?

(b)   What are the causes of the observed symptoms? Suggest why this condition, if prolonged, leads to brain damage.

Answer Secretion of insulin by the pancreas signals the liver that blood glucose levels (from dietary intake) are high. Circulating insulin stimulates the uptake and utilization of glucose by the liver. Once absorbed by liver cells, glucose is rapidly phosphorylated and converted into glycogen or broken down to pyruvate. In addition, the insulin signal decreases the rate of breakdown of amino acids and fatty acids, since an adequate supply of fuel is available. When blood glucose levels return to normal, insulin secretion decreases, and homeostasis is restored.

(a)   Excessive secretion of insulin promotes the excessive utilization of blood glucose by the liver, which leads to hypoglycemia (low blood glucose). To make matters worse, the high insulin level also shuts down amino acid and fatty acid catabolism.

(b)   Patients have little circulating fuel available to generate ATP for normal functions. Since glucose is the main source of fuel for the brain, decreased glucose levels cause neurological symptoms and prolonged hyperinsulinism leads to brain damage.

15. *Thermogenesis Caused by Thyroid Hormones* Thyroid hormones are intimately involved in regulating the basal metabolic rate. Liver tissue of animals given excess thyroxine shows an increased rate of O_2 consumption and increased heat output (thermogenesis), but the ATP concentration in the tissue is normal. Different explanations have been offered for the thermogenic effect of thyroxine. One is that excess thyroid hormone causes uncoupling of oxidative phosphorylation in mitochondria. How could such an effect account for the observations? Another explanation suggests that the thermogenesis is due to an increased rate of ATP utilization by the thyroid-stimulated tissue. Is this a reasonable explanation? Why?

Answer The observations are consistent with the thesis that thyroxine acts as an uncoupler of oxidative phosphorylation. Uncouplers lower the P/O ratio of tissues, and thus the tissue must increase respiration to meet the normal ATP demands. The observed thermogenesis could also be due to the increased rate of ATP utilization, since the increased ATP demands are met by increased oxidative phosphorylation and thus respiration. Despite much research, the details of how thyroid hormones regulate the rate of aerobic metabolism remain a mystery.

16. *Function of Prohormones* What are the possible advantages in the synthesis of hormones as prohormones or preprohormones?

Answer Several of the polypeptide hormones (e.g., insulin) are synthesized as inactive prohormones consisting of larger polypeptide chains than the active hormones themselves. One advantage of a prohormone is that it can be stored in quantity in secretory granules and activation of this store can occur rapidly by enzymatic cleavage (more rapidly than protein synthesis could occur) in response to an appropriate signal.

17. *Action of Aminophylline*   Aminophylline, a purine derivative resembling theophylline of tea, is often administered together with epinephrine to individuals with acute asthma.   What is the purpose and biochemical basis for this treatment?

> *Answer*   Epinephrine relaxes the smooth muscle surrounding the bronchioles of the lungs; it acts by stimulating the formation of cAMP in target cells.   Cyclic AMP is inactivated, however, by hydrolysis to AMP, catalyzed by phosphodiesterase.   Since this latter enzyme is inhibited by the purine derivatives caffeine, theophylline, and aminophylline, administering these drugs prolongs and intensifies the activity of epinephrine by decreasing the rate of breakdown of cAMP.

CHAPTER **23** **Genes and Chromosomes**

1. *How Long Is the Ribonuclease Gene?* What is the minimum number of nucleotide pairs in the gene for pancreatic ribonuclease (124 amino acids long)? Suggest a reason why the number of nucleotide pairs in the gene might be much larger than your answer.

> *Answer* Each gene encodes a polypeptide or other stable product, such as a tRNA or rRNA (see Fig. 23-6). Each amino acid is encoded by a triplet of nucleotides called a codon. A polypeptide of 124 amino acids is therefore encoded by (124 amino acids) (3 nucleotide pairs per amino acid) = 372 nucleotide pairs. The gene could be larger than this because of introns interrupting the exons and because additional amino acids may be encoded in a signal sequence required for secretion of ribonuclease from the pancreas.

2. *Packaging of DNA into a Virus* The DNA of bacteriophage T2 has a molecular weight of 120×10^6. The head of the T2 phage is about 210 nm long. Assuming the molecular weight of a nucleotide pair is 650, calculate the length of T2 DNA and compare it with the length of the T2 head. Your answer will show the necessity of very compact packaging of DNA in viruses (see Fig. 23-1).

> *Answer* The T2 DNA molecule has $(120 \times 10^6)/650 = 184{,}615$ nucleotide pairs; this is in close agreement with the value of 182,000 nucleotide pairs in Table 23-1. Recall from Chapter 12 that a nucleotide pair occupies 0.36 nm, so the length of the T2 DNA molecule is
>
> (184,615 nucleotide pairs)(0.36 nm per nucleotide pair) = 66,461 nm ≈ 66,000 nm
>
> Thus the DNA molecule is (66,000 nm)/210 nm = 314 ≈ 300 times longer than the T2 phage head.

3. *The DNA of Phage M13* Bacteriophage M13 DNA has the following base composition: A, 23%; T, 36%; G, 21%; C, 20%. What does this information tell us about the DNA of this phage?

> *Answer* The complementarity between A and T, and between G and C, in the two strands of duplex DNA explains Chargaff's rules that the sum of pyrimidine nucleotides equals that of purine nucleotides in DNAs from (virtually) all species: A = T, G = C, and A + G = C + T for duplex DNA. In M13 DNA, the percentage of A (23%) does not equal that of T (36%), nor does that of G (21%) equal that of C (20%). A + G = 44%, whereas C + T = 56%. This lack of equality between purine and pyrimidine nucleotides shows that M13 DNA is *not* double stranded, because the relationships expected from complementarity between the two strands of duplex DNA are not seen.

4. *Base Composition of ϕX174* **DNA** Bacteriophage ϕX174 DNA occurs in two forms, single-stranded in the isolated virion and double-stranded during viral replication in the host cell. Would you expect them to have the same base composition? Give your reasons.

> *Answer* No. The single-stranded viral form has a certain base composition. When it is copied into the duplex (or replicative) form, a T is incorporated into the new strand for every A in the viral strand, and so forth, following the rules of complementarity. Thus the duplex will have a different base composition from the single-stranded form. As a simple example, see how the base composition differs for a short sequence of DNA.
> Single stranded:
>
>> AGGGCTAAGC   A = 30%, G = 40%, C = 20%, T = 10%
>
> versus the double-stranded form:
>
>> AGGGCTAAGC   A = 20%, G = 30%, C = 30%, T = 20%
>> TCCCGATTCG

5. *Size of Eukaryotic Genes* An enzyme present in rat liver has a polypeptide chain of 192 amino acid residues. It is coded for by a gene having 1,440 nucleotide pairs. Explain the relationship between the number of amino acid residues in this enzyme and the number of nucleotide pairs in its gene.

> *Answer* Each amino acid is encoded by a triplet of three nucleotide pairs, so the 192 amino acids are encoded by 576 nucleotide pairs. The gene is in fact longer (1,440 nucleotide pairs). The additional 864 nucleotide pairs could be in introns (noncoding DNA, interrupting a coding segment) or they could code for a signal sequence (or leader peptide). In addition, as will be discussed in Chapter 25, eukaryotic mRNAs have untranslated segments before and after the portion coding for the polypeptide chain; these also contribute to the "extra" size of genes.

6. *DNA Supercoiling* A covalently closed circular DNA molecule has an *Lk* of 500 when it is relaxed. Approximately how many base pairs are in this DNA? How will the linking number be altered (increase, decrease, no change, become undefined) if (a) a protein complex is bound to form a nucleosome, (b) one DNA strand is broken, (c) DNA gyrase is added with ATP, or (d) the double helix is denatured (base pairs are separated) by heat?

> *Answer* In relaxed DNA, the linking number (*Lk*) is equivalent to the number of turns in the DNA helix. *Lk* is a topological property, and thus does not vary when duplex DNA is twisted or deformed, as long as both DNA strands remain intact. It can change only if one or both strands are broken and rejoined. If a DNA strand remains broken, the molecule is no longer topologically constrained (the strands can unravel) and *Lk* is undefined.
>
> The *Lk* of relaxed DNA is equivalent to the number of turns of DNA, and there are about 10.5 base pairs per turn of relaxed B-form DNA. Thus the DNA has approximately (500 turns)(10.5 bp/turn) = 5,250 base pairs.
>
> (a)   No change; the DNA strands are not cleaved and rejoined.
>
> (b)   Become undefined; one of the strands remains broken.
>
> (c)   Decrease; in the presence of ATP, gyrase (a type 2 topoisomerase that introduces negative supercoils) will underwind the DNA.
>
> (d)   No change; again, the DNA strands are not broken and rejoined.

7. **DNA Structure**   Explain how the underwinding of a B-DNA helix might facilitate or stabilize the formation of Z-DNA.

> **Answer**   The shift from B- to Z-DNA requires a shift from a right-handed helix to a left-handed helix. You can visualize the underwinding as aiding in the transition by providing a local region of unwinding that allows it to shift to Z-DNA. Second, the left-handed helix of Z-DNA has a negative Lk, and underwinding the DNA lowers the Lk. Third, the underwinding essentially stores some free energy within the DNA that can be used to facilitate the B to Z transition.

8. **Chromatin**   One of the important early pieces of evidence that helped define the structure of the nucleosome is illustrated by the agarose gel shown on p. 813 of the text, in which the thick bands represent DNA. It was generated by treating chromatin briefly with an enzyme that degrades DNA, then removing all protein and subjecting the purified DNA to electrophoresis. Numbers at the side of the gel denote the position to which a linear DNA of the indicated size (in base pairs) would migrate. What does this gel tell you about chromatin structure? Why are the DNA bands thick and spread out rather than sharp?

> **Answer**   The bands have a periodicity of about 200 nucleotide pairs, or base pairs, bp (200, 400, 600, etc.), showing that the chromatin is protected from nuclease digestion at regular intervals of 200 bp. This suggests that the nucleosomal cores (146 bp) were providing the protection, which was verified in numerous subsequent investigations. Thus the nucleosomes themselves are in a fairly regular array, occurring about once every 200 bp. The nuclease is cutting between the nucleosome cores (in spacer regions of about 60 bp), but it has not digested to completion. Some bands correspond to the DNA from single nucleosomes (200 bp), others to two nucleosomes (400 bp), and so forth. If the nucleosomes had been randomly distributed in the chromatin, a large number of differently sized DNA fragments would have been generated by the nuclease cleavage, and a heterogeneous population of DNA fragments would have smeared through the gel.
>
> The bands are thick because the spacer is fairly long (60 bp in this example) relative to the size of the nucleosomal core (146 bp). The nuclease can cut essentially anywhere in the spacer, so the band corresponding to, for example, mononucleosomes, has DNAs ranging from 146 to 206 bp.

CHAPTER 24 DNA Metabolism

1. **Conclusions from the Meselson-Stahl Experiment** The Meselson-Stahl experiment proved that DNA undergoes semiconservative replication in *E. coli*. In the "dispersive" model of DNA replication, the parent DNA strands are cleaved into pieces of random size and are then joined with pieces of the newly replicated DNA to yield daughter duplexes in which, in the Meselson-Stahl experiment, both strands would contain random segments of both heavy and light DNA. Explain how the results of the Meselson-Stahl experiment ruled out such a model.

   Answer If random, dispersive replication had taken place, the density of the first-generation DNA would have been the same as actually observed, a single band midway between heavy and light DNA. In the second generation all of the DNA would again have had the same density and would have appeared as a single band, midway between that observed in the first generation and that of light DNA. In fact, two bands were observed in the Meselson-Stahl experiment.

2. **Number of Turns in the E. coli Chromosome** How many turns must be unwound during replication of the *E. coli* chromosome? The chromosome contains about 4.7×10^6 base pairs.

   Answer During DNA replication, the complementary strands must unwind completely to allow the synthesis of a new strand on each template. Since there are 10.5 bp/turn in B-DNA, the number of helical turns = $\dfrac{\text{number of base pairs}}{\text{number of base pairs per helical turn}}$

   $\dfrac{4.7 \times 10^6 \text{ bp}}{10.5 \text{ bp/turn}}$ = 4.48×10^5 turns $\approx$ 4.5×10^5 turns

3. **Replication Time in E. coli** From the data in this chapter, how long would it take to replicate the *E. coli* chromosome at 37 °C, if two replication forks start from the origin? Under some conditions *E. coli* cells can divide every 20 min. Can you suggest how this is possible?

   Answer Chromosomal DNA replication in *E. coli* starts at a fixed origin and proceeds bidirectionally. Each replication fork travels $(4.7 \times 10^6)/2 = 2.35 \times 10^6$ nucleotide pairs during replication. From Table 24-1, the polymerization rate of DNA pol III is from 250 to 1000 nucleotides/s. If we use the highest number, the time required for the completion of DNA synthesis in each replication fork is $(2.35 \times 10^6)/1000 = 2350$ s = 39 min.

   One possible explanation is that replication of an *E. coli* chromosome starts from two origins, each proceeding bidirectionally to yield four replication forks. In this mode, it would take 19.5 min to complete the replication of the chromosome.

There is, however, only one replication origin in the *E. coli* chromosome. Thus, an alternative possibility is that a new round of replication begins before the previous one is completed: for cells dividing every 20 min, a replicative cycle is initiated every 20 min, and each daughter cell would receive a chromosome that is half-replicated. This latter mode has in fact been experimentally verified.

4. *Base Composition of DNAs Made from Single-Stranded Templates* Determine the base composition you might expect in the total DNA synthesized by DNA polymerase on templates provided by an equimolar mixture of the two complementary strands of circular bacteriophage φX174 DNA. The base composition of one strand is A, 24.7%; G, 24.1%; C, 18.5%; and T, 32.7%. What assumption is necessary to answer this problem?

   Answer The sequence of one strand of duplex DNA is complementary to that of the other strand, as determined by Watson-Crick base pairing (A with T, and G with C). The DNA strand made from the given template strand has A, 32.7%; G, 18.5%; C, 24.1%; T, 24.7%. The DNA strand made from the complementary template strand has A, 24.7%; G, 24.1%; C, 18.5%; T, 32.7%. Thus the composition of the *total* DNA synthesized is A, 28.7%; G, 21.3%; C, 21.3%; T, 28.7%. It is assumed that both template strands are completely replicated.

5. *Okazaki Fragments* In the replication of the *E. coli* chromosome, about how many Okazaki fragments would be formed? What factors guarantee that the numerous Okazaki fragments are assembled in the correct order in the new DNA?

   Answer About 2,350 to 4,700 Okazaki fragments (4.7×10^6 bp/2000 bp and 4.7×10^6 bp/1000 bp, respectively) are formed by DNA polymerase III from an RNA primer and a DNA template. Because the Okazaki fragments in *E. coli* are 1,000 to 2,000 bases long, they are firmly bound to the template strand by base pairing. Each fragment is quickly joined to the lagging strand by the successive action of DNA polymerase I and DNA ligase, thus preserving the correct order of the fragments. A mixed pool of different Okazaki fragments, detached from their template, does *not* form during normal replication.

6. *Leading and Lagging Strands* List and compare the precursors and enzymes needed to make the leading versus lagging strands during DNA replication in *E. coli*.

   Answer This discussion refers to the replication of DNA in *E. coli*. The *leading strand* is produced by continuous replication of the DNA template strand in the 5'→3' direction. The precursors required are dATP, dGTP, dCTP, and dTTP, which serve as the source of nucleotides for the new DNA strand. For some of the steps, ATP is needed as the energy source. A template DNA strand and a priming DNA strand are also required. The required enzymes are: *DNA helicase,* which unwinds short segments of the DNA helix just ahead of the replicating fork; it requires ATP. *Single-strand DNA-binding proteins*, which bind tightly to the separated strands to prevent base pairing while the templates are being replicated. *DNA gyrase,* a topoisomerase, which permits swiveling of the DNA, twisting it in a direction that favors unwinding of the strands at the replication fork; it requires ATP. *DNA polymerase III,* which carries out the elongation by addition of nucleotide units; the cofactors Mg^{2+} and Zn^{2+} are required. *Pyrophosphatase,* which hydrolyzes the PP_i released as each new nucleotide unit is added; this hydrolysis helps "pull" the reaction in the forward direction.

The *lagging strand*, which is synthesized in the form of Okazaki fragments that are then spliced together, requires dATP, dGTP, dCTP, and dTTP as the source of nucleotides in the new DNA strand. *UTP, *ATP, *CTP, and *GTP are required for formation of the RNA primer that starts each Okazaki fragment. The required enzymes, *in addition* to those listed for the leading strand, are: *Primase,* which constructs a short RNA primer, complementary to the DNA template, to initiate the Okazaki fragment. *DNA polymerase I*, which removes the RNA primer (exonuclease activity), replacing each NMP unit with a dNMP unit (polymerase activity); cofactors required are Zn^{2+} and Mg^{2+}. *DNA ligase*, which carries out the final step of splicing the new fragment to the lagging strand (this involves *NAD$^+$ as energy source).

The precursors and enzymes marked with an asterisk are required *only* for the lagging strand.

7. *Fidelity of Replication of DNA* What factors participate in ensuring the fidelity of replication during the synthesis of the leading strand of a new DNA? Would you expect the lagging strand to be made with the same fidelity as the leading strand? Give reasons for your answers.

   Answer Fidelity of replication is ensured by Watson-Crick base pairing between the template and leading strand, and proofreading and removal of wrongly inserted nucleotides by the 3'-exonuclease activity of DNA polymerase III. The same fidelity would be expected in the lagging strand—maybe. The factors ensuring fidelity of replication are operative in both the leading and the lagging strands. However, the greater number of distinct chemical operations involved in making the lagging strand might provide a greater opportunity for errors to arise.

8. *DNA Repair Mechanisms* Vertebrate and plant cells often methylate cytosine in DNA to form 5-methylcytosine (see Fig. 12-5a). In these same cells, there is a specialized repair system that recognizes G-T mismatches and repairs them to $G \equiv C$ base pairs. Rationalize this repair system in terms of the presence of 5-methylcytosine in the DNA.

   Answer This problem is very similar to Problem 7 of Chapter 12. Spontaneous deamination of 5-methylcytosine produces thymine, and thus a G-T mismatched pair. Such G-T mismatches are among the most common mismatches in the DNA of eukaryotes. The specialized repair system restores the $G \equiv C$ pair.

9. *Holliday Intermediates* How are the Holliday intermediates formed in homologous genetic recombination and in site-specific recombination different?

   Answer During homologous genetic recombination, a Holliday intermediate may be formed almost anywhere within the two paired, homologous chromosomes. Once formed, the branch point of the intermediate may move extensively by branch migration. In site-specific recombination, the Holliday intermediate is formed between two specific sites, and branch migration is generally restricted by heterologous sequences on either side of the recombination sites.

10. **DNA Recombination**   A circular DNA molecule is converted to two smaller circles by an enzyme or enzymes in a crude cellular extract. What types of recombination could account for this reaction, and what else must you know to determine which type it is?

> **Answer** Homologous or site-specific recombination; the two can generally be distinguished by determining the nature of the sequences that have been recombined and the location of the crossover event. If the recombination has occurred between two long (>50 base pair) homologous sequences, and the crossover could occur anywhere within the sequences, it is homologous recombination. If the crossover has occurred between two short homologous sequences and always involved the same phosphodiester bond within those sequences, it is site-specific recombination.

CHAPTER 25   RNA Metabolism

1. **RNA Polymerase** How long would it take for the *E. coli* RNA polymerase to synthesize the primary transcript for *E. coli* rRNAs (6500 bases)?

    Answer Elongation of the RNA transcript in *E. coli* proceeds at about 50 nucleotides per second. Thus the time required to produce the primary transcript is

    $$\frac{6500 \text{ nucleotides}}{50 \text{ nucleotides/s}} = 130 \text{ s} \approx 2 \text{ min}$$

2. **Error Correction by RNA Polymerases** DNA polymerases are capable of editing and error correction, but RNA polymerases do not appear to have this capacity. Given that a single base error in either replication or transcription can lead to an error in protein synthesis, can you give a possible biological explanation for this striking difference?

    Answer Since many RNA molecules are made from each gene, an error in any one molecule will lead to only a small fraction of protein with an incorrect amino acid. The incorrect protein will probably be degraded fairly quickly. The error in the mRNA will not be propagated in subsequent generations of cells because the mRNA itself is degraded. For DNA replication, however, errors would be transmitted to the next generation of cells.

3. **The Rate of Transcription** From what you know of the rate at which *E. coli* RNA polymerase synthesizes RNA, predict how far the transcription "bubble" formed by RNA polymerase will move along the DNA in 10 s.

    Answer Elongation of the RNA transcript proceeds at about 50 nucleotides per second. Thus in 10 s the bubble travels

    $$(10 \text{ s})(50 \text{ nucleotides/s}) = 500 \text{ nucleotides}$$

4.   *RNA Posttranscriptional Processing* Predict the likely effects of a mutation in the sequence (5′)AAUAAA in a eukaryotic mRNA transcript.

>   ***Answer*** One of the key signals for cleavage and 3′ polyadenylation is the sequence AAUAAA. After RNA polymerase II has transcribed beyond this sequence, an endonuclease (uncharacterized at this time) cleaves the primary transcript at a position about 25 to 30 nucleotides 3′ to the AAUAAA. The enzyme polyadenylate polymerase then adds a string of 20 to 250 A residues to the free 3′ end, generating the 3′ poly(A) tail. The mutation would prevent cleavage and polyadenylation at the usual site. If the transcript is not polyadenylated it will be quite unstable, the steady-state levels of mRNA will be very low, and little or no protein product will be made.

5.   *Coding vs. Template Strands* The RNA genome of phage Qβ is the nontemplate or (+) strand, and when introduced into the cell it functions as an mRNA. Suppose the RNA replicase of phage Qβ synthesized primarily (-) strand RNA and uniquely incorporated it into the virus particles, rather than (+) strands. What would be the fate of the (-) strands when they entered a new cell? What enzyme would such a (-) strand virus need to include in the virus particle to successfully invade a host cell?

>   ***Answer*** A (-) strand RNA does not encode proteins, and by itself would not cause a productive infection. However, if an RNA-dependent RNA polymerase were included in the viral particle, it could copy the (-) strand to form a (+) strand after entry into the host cell. The (+) strand would serve as mRNA for synthesis of viral proteins, leading to a productive infection.

6.   *The Chemistry of Nucleic Acid Biosynthesis* Describe three properties common to the reactions catalyzed by DNA polymerase, RNA polymerase, reverse transcriptase, and RNA replicase.

>   ***Answer*** These enzymes have at least four properties in common.
>
>   (1)   All are template directed, synthesizing a sequence complementary to the template.
>
>   (2)   Synthesis occurs in a 5′ → 3′ direction.
>
>   (3)   All catalyze the addition of a nucleotide by the formation of a phosphodiester bond.
>
>   (4)   All use (deoxy)ribonucleoside triphosphates as substrate, and release pyrophosphate as a product.
>
>   The enzyme polynucleotide phosphorylase differs from these polymerase in points 1 and 4 (p. 880). Polynucleotide phosphorylase does not use a template, but rather adds ribonucleotides to an RNA in a highly reversible reaction. The substrates (in the direction of synthesis) are ribonucleoside diphosphates, which are added with the release of phosphate as a product. In the cell, this enzyme probably catalyzes the reverse reaction to degrade RNAs.

7. **RNA Splicing** What is the minimum number of transesterification reactions needed to splice an intron from an mRNA transcript? Why?

>   **Answer** A minimum of two transesterification steps are required. The mechanism for removal of introns from pre-mRNAs involves the formation of a lariat intermediate after the reaction is initiated. Each of the cleavage and rejoining reactions is a transesterification, in which a new phosphodiester bond is formed for every one that is broken. The first step is initiated by the attack of the 2'-hydroxyl group of an A residue within the intron on the bond linking the 3' end of the first exon with the 5' end of the intron. This generates a 3'-hydroxyl on the nucleotide at the 3' end of the first exon, and effectively takes the intron out of the series of transesterifications by forming a lariat structure (see Fig. 25-14). The 3' nucleotide of the first intron can then link to the first nucleotide of the second exon, again by a transesterification. The result of this second step is the union of the first and second exons, with the intron released as a lariat intermediate.

8. **Telomerase** Assuming that the RNA component of telomerase is fixed within the protein structure, in what respect might the active site of this enzyme differ from the active site of reverse transcriptases, RNA polymerases, and DNA polymerases? (Hint: The latter three enzymes add one nucleotide at a time.)

>   **Answer** Telomerase is a ribonucleoprotein that catalyzes repetitive addition of T_xG_y to the ends of chromosomes. The added string of deoxyribonucleotides is complementary to a portion of the RNA component of the enzyme, which serves as the template for the addition of the deoxyribonucleotides. Telomerase may be able to add all the nucleotides complementary to its internal template while at one position at the end of the chromosome. For example, in the situation illustrated in Fig. 25-34, the telomerase could add GGGTTTTG to the end of the DNA strand, then translocate to repeat the synthesis of GGGTTTTG. Other polymerases, in contrast, add one nucleotide then translocate to the next position along the template and add another nucleotide.

9. **RNA Genomes** The RNA viruses have relatively small genomes. For example, the single-stranded RNAs of retroviruses have about 10,000 nucleotides and the Qβ RNA is only 4,220 nucleotides long. Given the properties of reverse transcriptase and RNA replicase described in this chapter, can you suggest a reason for the small size of these viral genomes?

>   **Answer** Neither reverse transcriptase nor RNA replicase has a proofreading function, and hence these enzymes are much more error-prone than DNA polymerases. The smaller the genomes, the fewer are the mutable sites, and thus it is less likely that a lethal mutational load will accumulate (i.e., that the RNA will sustain one or a group of mutations that inactivates the virus).

CHAPTER **26**   **Protein Metabolism**

1. *Messenger RNA Translation* Predict the amino acid sequences of peptides formed by ribosomes in response to the following mRNAs, assuming that the initial codon is the first three bases in each sequence.

    (a) GGUCAGUCGCUCCUGAUU

    (b) UUGGAUGCGCCAUAAUUUGCU

    (c) CAUGAUGCCUGUUGCUAC

    (d) AUGGACGAA

    Answer The genetic code is nonoverlapping, unpunctuated, and in triplet (see Fig. 26-7).

    (a) Gly-Gln-Ser-Leu-Leu-Ile

    (b) Leu-Asp-Ala-Pro

    (c) His-Asp-Ala-Cys-Cys-Tyr

    (d) Met-Asp-Glu in eukaryotes; fMet-Asp-Glu in prokaryotes

2. *How Many mRNAs Can Specify One Amino Acid Sequence?* Write all the possible mRNA sequences that can code for the simple tripeptide segment Leu-Met-Tyr. Your answer will give you some idea as to the number of possible mRNAs that can code for one polypeptide.

    Answer The genetic code is degenerate, meaning that a given amino acid may be specified by more than one codon (see Table 26-4). Leu is specified by six different codons: UUA, UUG, CUU, CUC, CUA, CUG. Met, when not used as an initiation codon, is specified by AUG; Tyr is specified by two codons: UAC, UAU. Thus, there are 6 x 1 x 2 = 12 possible mRNA sequences that can code for a tripeptide segment Leu-Met-Tyr:

    UUA AUG UAU, UUG AUG UAU, CUU AUG UAU,
    CUC AUG UAU, CUA AUG UAU, CUG AUG UAU,
    UUA AUG UAC, UUG AUG UAC, CUU AUG UAC,
    CUC AUG UAC, CUA AUG UAC, CUG AUG UAC

3. *Can the Base Sequence of an mRNA Be Predicted from the Amino Acid Sequences of Its Polypeptide Product?* A given sequence of bases in an mRNA will code for one and only one sequence of amino acids in a polypeptide, if the reading frame is specified. From a given sequence of amino acid residues in a protein such as cytochrome *c*, can we predict the base sequence of the unique mRNA that coded for it? Give reasons for your answer.

Answer No; because nearly all the amino acids have more than one codon, any given polypeptide can be coded for by a number of different base sequences (see Problem 2). However, because some amino acids are encoded by only one codon and those with multiple codons often share the same nucleotide at two of the three positions, *certain parts* of the mRNA sequence encoding a protein of known amino acid sequence can be predicted with high certainty.

4. *Coding of a Polypeptide by Duplex DNA* The template strand of a sample of double-helical DNA contains the sequence

      (5′)CTTAACACCCCTGACTTCGCGCCGTCG

(a) What is the base sequence of mRNA that can be transcribed from this strand?

(b) What amino acid sequence could be coded by the mRNA base sequence in (a), starting from the 5′ end?

(c) Suppose the other (nontemplate) strand of this DNA sample is transcribed and translated. Will the resulting amino acid sequence be the same as in (b)? Explain the biological significance of your answer.

Answer The template strand serves as the template for RNA synthesis; the nontemplate strand is identical in sequence to the RNA transcribed from the gene, with U in place of T.

(a)   (5′)CGACGGCGCGAAGUCAGGGGUGUUAAG(3′)

(b)   Arg-Arg-Arg-Glu-Val-Arg-Gly-Val-Lys

(c)   No; the base sequence of mRNA transcribed from the nontemplate strand would be (5′)CUUAACACCCCUGACUUCGCGCCGUCG. This mRNA, when translated, would result in a different peptide from (b). The complementary antiparallel strands in double-helical DNA do not have the same base sequence in the 5′→3′ direction. RNA is transcribed from only one specific strand of duplex DNA. The RNA polymerase must therefore recognize and bind to the correct strand.

5. *Methionine Has Only One Codon* Methionine is one of the two amino acids having only one codon. Yet the single codon for methionine can specify both the initiating residue and interior Met residues of polypeptides synthesized by *E. coli*. Explain exactly how this is possible.

Answer There are two tRNAs for methionine: tRNA^{fMet}, the initiating tRNA, and tRNA^{Met}, which can insert Met in interior positions in a polypeptide. The tRNA^{fMet} reacts with Met to yield Met-tRNA^{fMet}, promoted by methionine aminoacyl-tRNA synthetase. The amino group of its Met residue is then formylated by N^{10}-formyltetrahydrofolate to yield fMet-tRNA^{fMet}. Free Met or Met-tRNA^{Met} cannot be formylated. Only fMet-tRNA^{fMet} is recognized by the initiation factor IF-2 and is aligned with the initiating AUG positioned at the ribosomal P site in the initiation complex. AUG codons in the interior of the mRNA are eventually positioned at the ribosomal A site and can bind and incorporate only Met-tRNA^{Met}.

6. *Synthetic mRNAs* How would you make a polyribonucleotide that could serve as an mRNA coding predominantly for many Phe residues and a small number of Leu and Ser residues? What other amino acid(s) would be coded for by this polyribonucleotide but in smaller amounts?

> *Answer* Polynucleotide phosphorylase is template-independent and does not require a primer. The base composition of the RNA formed by this enzyme reflects the relative concentrations of the nucleoside 5′-diphosphates in the reaction mixture. To prepare the required polyribonucleotide, allow polynucleotide phosphorylase to act on a mixture of UDP and CDP in which UDP has, say, five times the concentration of CDP. The result would be a synthetic RNA polymer with many UUU triplets (coding for Phe), a smaller number of UUC (also Phe), UCU (Ser), and CUU (Leu), and a yet smaller number of UCC (also Ser), CUC (also Leu), and CCU (Pro).

7. *The Direct Energy Cost of Protein Biosynthesis* Determine the minimum energy cost, in terms of high-energy phosphate groups expended, required for the biosynthesis of the β-globin chain of hemoglobin (146 residues), starting from a pool including all necessary amino acids, ATP, and GTP. Compare your answer with the direct energy cost of the biosynthesis of a linear glycogen chain of 146 glucose residues in ($\alpha1\rightarrow4$) linkage, starting from a pool including glucose, UTP, and ATP (Chapter 19). From your data, what is the *extra* energy cost of imparting the genetic information inherent in the β-globin molecule?

> *Answer* The number of high-energy phosphate groups required for the synthesis of a *polypeptide* with n residues is:
>
> $2n$ for the charging of tRNA (ATP $\longrightarrow$ AMP + PP$_i$; PP$_i \longrightarrow$ 2P$_i$)
>
> 1 for initiation (GTP $\longrightarrow$ GDP + P$_i$)
>
> n - 1 for the formation of n - 1 peptide bonds (GTP $\longrightarrow$ GDP + P$_i$)
>
> n - 1 for the n - 1 translocation steps (GTP $\longrightarrow$ GDP + P$_i$)
>
> The total number for the polypeptide = $4n$ - 1.
>
> The number of high-energy phosphate groups required for the synthesis of a linear *glycogen* chain of n glucose residues is:
>
> n for the phosphorylation of glucose (ATP $\longrightarrow$ ADP)
>
> None for the conversion of glucose-6-phosphate to glucose-1-phosphate
>
> $2n$ for the activation of glucose-1-phosphate to UDP-glucose (UTP $\longrightarrow$ PP$_i \longrightarrow$ 2P$_i$)
>
> -n (i.e., n generated) for the formation of the polymer from UDP-glucose (UDP-glucose $\longrightarrow$ UDP)
>
> The total number for the glycogen chain = $2n$.
>
> Thus the total number of high-energy phosphate groups required for the synthesis of one molecule of β-globin is (4 x 146) - 1 = 583. The total number for the synthesis of a linear chain of 146 glucose residues is 2 x 146 = 292. The extra energy cost for the synthesis of β-globin is 583 - 292 = 291 high-energy phosphate groups; this reflects the cost of the information contained in the protein.

8. **Indirect Costs of Protein Synthesis** In addition to the direct energy cost for the synthesis of a protein, as developed in Problem 7, there are indirect energy costs—those required for the cell to make the necessary biocatalysts for protein synthesis. Contrast the relative magnitude of the indirect costs to a eukaryotic cell of the biosynthesis of linear ($\alpha 1 \rightarrow 4$) glycogen chains versus the indirect costs of the biosynthesis of polypeptides. (Compare the enzymatic machinery used to synthesize proteins and glycogen.)

> **Answer** At least 20 aminoacyl-tRNA synthetases (activating enzymes), 70 ribosomal proteins, 4 rRNAs, 32 or more tRNAs, an mRNA, and 10 or more auxiliary enzymes must be made by the eukaryotic cell in order to synthesize a protein from amino acids. Synthesis of these proteins and RNA molecules is energetically expensive. In contrast, the synthesis of an ($\alpha 1 \rightarrow 4$) chain of glycogen from glucose requires only four or five enzymes (see Chapter 19).

9. **Predicting Anticodons from Codons** Most amino acids have more than one codon and will be attached to more than one tRNA, each with a different anticodon. Write all possible anticodons for the four codons for glycine: (5′)GGU, GGC, GGA, and GGG.

   (a) From your answer, which of the positions in the anticodons are primary determinants of their codon specificity in the case of glycine?

   (b) Which of these anticodon-codon pairings have a wobbly base pair?

   (c) In which of the anticodon-codon pairings do all three positions exhibit strong Watson-Crick hydrogen bonding?

> **Answer** All the anticodons for the four Gly codons have the sequence (5′)XCC. The first position of each anticodon is determined by the specific codon it interacts with, and by the wobble hypothesis (see Table 26-5). For example, the wobble position of the codon GGU is U, which can be recognized by either A, or G, or I. Thus, this codon has three possible anticodons: ACC, GCC, and ICC. By the same token, the anticodons for the GGC codon are GCC and ICC; the anticodons for the GGA codon are UCC and ICC; and the anticodons for the GGG codon are CCC and UCC.
>
> (a) The 3′ and the middle position. The 5′ position is the wobble position.
>
> (b) Anticodons GCC, ICC, and UCC each recognize more than one codon by virtue of wobble base pairing. Anticodons ACC and CCC each only recognize one codon and thus are not involved in wobble base pairing.
>
> (c) The pairing of anticodons ACC and CCC with their respective codons involves Watson-Crick base pairing in all three positions, A pairing with U, and C pairing with G.

10. **The Effect of Single-Base Changes on Amino Acid Sequence** Much important confirmatory evidence on the genetic code has come from the nature of single-residue changes in the amino acid sequence of mutant proteins. Which of the following single-residue amino acid replacements would be consistent with the genetic code? Which cannot be the result of single-base mutations? Why?

   (a) Phe → Leu
   (b) Lys → Ala
   (c) Ala → Thr
   (d) Phe → Lys
   (e) Ile → Leu
   (f) His → Glu
   (g) Pro → Ser

Answer For each part of this problem "yes" indicates that the replacement is consistent with a single-base change. The various codons for each amino acid are listed. The single-base changes (where they exist) that would cause the mutation are underlined.

(a) Yes.

Phe: UUU UUC UUA UUG
Leu: CUU CUC CUA CUG

(b) No single-base mutation could convert the codons for Lys to the codons for Ala.

Lys: AAA AAG
Ala: GCU GCC GCA GCG

(c) Yes.

Ala: GCU GCC GCA GCG
Thr: ACU ACC ACA ACG

(d) No single-base mutation could convert the codons for Phe to the codons for Lys.

Phe: UUU UUC UUA UUG
Lys: AAA AAG

(e) Yes.

Ile: AUU AUC AUA
Leu: CUU CUC CUA

(f) No single-base mutation could convert the codons for His to the codons for Glu.

His: CAU CAC
Glu: GAA GAG

(g) Yes.

Pro: CCU CCC CCA CCG
Ser: UCU UCC UCA UCG

11. *The Basis of the Sickle-Cell Mutation* In sickle-cell hemoglobin there is a Val residue at position 6 of the β-globin chain, instead of the Glu residue found in this position in normal hemoglobin A. Can you predict what change took place in the DNA codon for glutamate to account for its replacement by valine?

Answer The two DNA codons for Glu are GAA and GAG, and the four DNA codons for Val are GTT, GTC, GTA, and GTG. A single-base change in GAA to form GTA or in GAG to form GTG could account for the Glu → Val replacement in sickle-cell hemoglobin. Much less likely are two-base changes from GAA to GTG, GTT, or GTC; and from GAG to GTA, GTT, or GTC.

12. *Importance of the "Second Genetic Code"* Some aminoacyl-tRNA synthetases do not bind the anticodon of their cognate tRNAs but instead use other structural features of the tRNAs to impart binding specificity. The tRNAs for alanine apparently fall into this category. Describe the consequences of a C → G mutation in the third position of the anticodon of tRNA^{Ala}. What other kinds of mutations might have similar effects? Mutations of these kinds are never found in natural populations of any organism. Why? (Hint: Consider what might happen both to individual proteins and to the organism as a whole.)

> *Answer* The only nucleotides of tRNA^{Ala} required for recognition by Ala-tRNA synthetase are those of the G^3-C^{70} base pair in the amino acid arm (see Fig. 26-21). Thus, changes in the anticodon of tRNA^{Ala} would not affect the specificity of the charging reaction by Ala-tRNA synthetase. However, depending on the changes, the codons recognized by mutant tRNA^{Ala} may specify amino acids other than alanine. There are four Ala codons, GCU, GCC, GCA, and GCG. The third position of each tRNA^{Ala} should be a C, because this position interacts with the third position of the Ala codons, which is a G in all four. Thus, changing the C in the third position of tRNA^{Ala} to a G would allow the mutant tRNA^{Ala} to recognize CCU, CCC, CCA, and CCG, all of which specify Pro. Thus Ala residues would be inserted at sites coding for Pro. The Ala ⟶ Pro replacement resulting from these mutations will render most of the proteins in the cell inactive, making these mutations lethal; hence, their effects would not be observed.

13. *Maintaining the Fidelity of Protein Synthesis* The chemical mechanisms used to avoid errors in protein synthesis are different from those used during DNA replication. DNA polymerases utilize a 3'→5' exonuclease proofreading activity to remove mispaired nucleotides incorrectly inserted into a growing DNA strand. There is no analogous proofreading function on ribosomes; and, in fact, the identity of amino acids attached to incoming tRNAs and added to the growing polypeptide is never checked. A proofreading step that hydrolyzed the last peptide bond formed when an incorrect amino acid was inserted into a growing polypeptide (analogous to the proofreading step of DNA polymerases) would actually be chemically impractical. Why? (Hint: Consider how the link between the growing polypeptide and the mRNA is maintained during the elongation phase of protein synthesis; see Figs. 26-27 and 26-28.)

> *Answer* The amino acid most recently added to a growing polypeptide chain is the only one covalently attached to a tRNA and hence is the only link between the polypeptide and the mRNA that is encoding it. A proofreading activity that severed this link would halt synthesis of the polypeptide and release it from the mRNA.

CHAPTER **27** **Regulation of Gene Expression**

1. *Negative Regulation* In the *lac* operon, describe the probable effect on gene expression of:

   (a) Mutations in the *lac* operator

   (b) Mutations in the *lac*I gene

   (c) Mutations in the promoter

   > **Answer** The *lac* operon is negatively regulated by a repressor, the product of the *lac*I gene. The *lac* repressor binds to a specific DNA sequence called the operator (*lac*O) and prevents efficient initiation of transcription by RNA polymerase from the promoter (*lac*P). An inducer (allolactose or an analog) binds to the repressor and prevents its binding to the operator, thereby releasing the repression and allowing transcription of the *lac* operon.

   > (a) Most mutations in the operator, the binding site for repressor, lead to lower affinity for the repressor and hence less binding. Thus these mutations allow continued transcription (and thus expression) of the *lac* operon even in the absence of inducer; this is referred to as constitutive expression.

   > (b) Mutations in the *lac*I gene that produce a repressor that cannot bind to the operator will lead to constitutive expression (no repression in the absence of inducer). Mutations that prevent binding of the repressor to the inducer without affecting the ability to bind to the operator lead to a noninducible phenotype.

   > (c) The *lac* promoter is not a particularly strong promoter. Mutations have been observed that either increase or decrease its efficiency of initiating transcription. Base substitutions that make the promoter sequence more similar to the consensus generate a stronger promoter (promoter "up" mutations), whereas those that make the promoter less similar to the consensus generate a weaker promoter (promoter "down" mutations). An "up" mutation would make the *lac* operon independent of positive regulation by the cAMP-CAP complex (when the operon is induced). A "down" mutation would not allow expression even in the derepressed state (presence of inducer) and hence would produce a noninducible phenotype.

2. *Effect of mRNA and Protein Stability on Regulation* An *E. coli* cell is growing in a solution with glucose as the sole carbon source. Tryptophan is suddenly added. The cells continue to grow, and divide every 30 min. Describe (qualitatively) how the amount of tryptophan synthase activity in the cell changes if:

   (a) The *trp* mRNA is stable (degraded slowly over many hours).

   (b) The *trp* mRNA is degraded rapidly, but tryptophan synthase is stable.

   (c) The *trp* mRNA and tryptophan synthase are both degraded rapidly.

Answer The mRNA from the *trp*EDCBA operon encodes several enzymes for Trp biosynthesis; the *trp*B and *trp*A genes encode tryptophan synthase. The complete *trp* mRNA is synthesized only when the concentration of Trp (actually that of charged Trp-tRNA) is low. This is the result of both repression and attenuation (see Problem 6). There is no strong regulation at translation, so when *trp* mRNA is present, tryptophan synthase will be produced. Although the regulation of Trp biosynthesis, like most biosynthetic pathways, is fine-tuned by feedback control, feedback inhibition by Trp is exerted at the branch point anthranilate synthase (the product of the *trp*E and *trp*D genes; see Fig. 27-21), and hence the activity of tryptophan synthase is not strongly affected by [Trp].

(a)   If the *trp* mRNA is stable relative to cell generation time, it will persist in the population of bacteria even after Trp has been added, and the enzyme tryptophan synthase will continue to be synthesized and active. In a simple model, one would expect to see the enzyme activity decrease roughly twofold per cell for each generation (30 min); that is, the activity is slowly diluted out by the increasing numbers of cells.

(b)   Again, if the enzyme is stable relative to the generation time, it will persist in the population, even after the addition of Trp, and be active.

(c)   If both mRNA and enzyme are unstable (degraded more rapidly than the cells divide), the attenuation of transcription of the *trp* operon caused by the addition of Trp to the medium will lead to an abrupt decrease in levels of *trp* mRNA and tryptophan synthase.

3.   *Functional Domains in Regulatory Proteins*   A biochemist replaces the DNA-binding domain of the yeast GAL4 protein with the DNA-binding domain from the λ repressor (CI) and finds that the engineered protein no longer functions as a transcriptional activator (it no longer regulates transcription of the *gal* operon in yeast). What might be done to the GAL4 DNA-binding site to make the engineered protein functional in activating *gal* operon transcription?

Answer Transcriptional activators have at least two domains that frequently function separately: the DNA-binding domain and the activation domain. The DNA-binding domain is required for the sequence-specific binding of the protein to DNA. A different portion of the protein is responsible for activation; this domain may directly contact the RNA polymerase or it may facilitate the action of coactivators or other proteins that stimulate transcription.

The biochemist has done part of a domain-swap experiment; the activation domain of GAL4 is fused to the DNA-binding domain of the λ repressor. This new hybrid will no longer recognize the GAL4 binding site in DNA (called UAS$_G$), since that DNA-binding domain is no longer present. However, replacement of UAS$_G$ with the λ operator (binding site for the λ repressor) in the *gal* operon should allow the λ repressor-GAL4 hybrid protein to function as a transcriptional activator of *gal* genes in yeast.

4. *Bacteriophage* λ Bacteria that become lysogenic for bacteriophage λ are immune to subsequent λ lytic infections. Why?

   Answer Lysogeny results from the integration of a λ prophage under conditions where the expression of all the λ genes except cI are repressed. Bacteria that are lysogenic for λ are already producing the CI protein, (α repressor). Subsequent infection by another λ phage results in the immediate binding of the λ repressor to the leftward and rightward operators of the incoming phage, thereby preventing transcription of any of its genes, including those required for lytic infection (see Fig. 27-27c). The newly introduced phage DNA also cannot integrate (no expression of *int* and *xis*), and it is eventually degraded.

5. *Regulation by Means of Recombination* In the phase variation system of *Salmonella*, what would happen to the cell if the Hin recombinase became more active and promoted recombination (the switch) several times in each cell generation?

   Answer *Salmonella* will switch expression between two different flagellin genes, *H1* and *H2*, about once every 1000 generations in order to evade the immune system of the host organism. This switch in expression is accomplished by a site-specific recombination system, requiring the action of the Hin recombinase on *hix* sites that flank the *H2* gene. *H2* is expressed and *H1* is repressed in one orientation of *hin* (the one in which the *H2* promoter is orientated toward the *H2* gene), whereas *H1* is expressed and *H2* is inactive in the other orientation (see Fig. 27-32).

   More rapid switching due to a more active Hin recombinase would lead to a mixed population of *Salmonella*, some with H1 flagellin and others with H2 flagellin. The host immune system, faced with both types of flagella, would mount an attack on both, greatly reducing the numbers of *Salmonella*. In other words, the protective advantage (to *Salmonella*) of phase variation would be lost.

6. *Transcription Attenuation* In the leader region of the *trp* mRNA, what would be the effect of:
   (a) Increasing the distance (number of bases) between the leader peptide gene and sequence 2?
   (b) Increasing the distance between sequences 2 and 3?
   (c) Removing sequence 4?

   Answer The *trp* operon is subject to regulation by both repression and attenuation. Attenuation depends on the tight coupling between transcription and translation in bacteria. When the Trp concentration is high, translation of the *trp* leader is completed and the ribosome blocks sequence 2. This allows the transcribed sequences 3 and 4 to form the stem and loop attenuator structure (see Fig. 27-23). Formation of the 3:4 loop, which resembles a rho-independent transcription terminator, terminates transcription of the *trp* operon before the structural genes (EDCBA) are transcribed, and the enzymes for Trp biosynthesis are not produced. When [Trp] is low, translation of the *trp* leader is stalled at two Trp codons. In this position, the ribosome does not cover sequence 2, and sequence 2 can base-pair with sequence 3 in an alternative secondary structure. Formation of the 2:3 stem and loop precludes formation of the 3:4 attenuator loop, and transcription proceeds through the *trp*EDCBA genes. Thus when [Trp] is low the biosynthetic genes are expressed and more Trp is synthesized.

(a)   Increasing the distance between sequence 1 (encoding the *trp* leader peptide) and sequence 2 will decrease attenuation when [Trp] is high. In this situation, the ribosome, after completing translation of the leader, will not cover sequence 2. Hence the 2:3 stem and loop can form, preventing formation of the 3:4 structure and thereby losing the normal attenuation (visualize the top part of Fig. 27-23a with a greater distance between sequences 1 and 2).

(b)   A large increase in the distance between sequences 2 and 3 could disfavor formation of the 2:3 stem and loop and hence formation of the 3:4 attenuator structure. Thus at low [Trp], even though the ribosome has stalled, the 2:3 loop would not form, allowing formation of the 3:4 attenuator structure, resulting in a decreased *trp* operon expression (due to attenuation).

(c)   Because sequence 4 is required to form the 3:4 attenuator stem and loop, in its absence no attenuation would occur.

7.   *Specific DNA Binding by Regulatory Proteins*   A typical prokaryotic repressor protein discriminates between its specific DNA-binding site (operator) and nonspecific DNA by a factor of 10^5 to 10^6. About ten molecules of the repressor per cell are sufficient to ensure a high level of repression. Assume that a very similar repressor existed in a human cell and had a similar specificity for its binding site. How many copies of the repressor would be required per cell to elicit a level of repression similar to that seen in the prokaryotic cell? (Hint: The *E. coli* genome contains about 4.7 million base pairs and the human haploid genome contains about 2.4 billion base pairs).

Answer   The discrete DNA-binding domains of transcriptional regulatory proteins form specific complexes with defined sequences of DNA. Their affinity for these defined sequences is about 10^5 to 10^6 greater than their affinity for other sequences.

Using the example of the *lac* repressor, the binding site (operator) is 22 base pairs (bp) long. Ten molecules of the *lac* repressor are sufficient to keep this operator in a bound state even in the context of 4.7×10^6 bp of nonspecific DNA (the rest of the *E. coli* genome). This amounts to finding one specific site in a sea of $(4.7 \times 10^6 \text{ bp})/(22 \text{ bp}) = 2.1 \times 10^5$ nonspecific sites. For the hypothetical repressor in a human cell, let's use the same size binding site (22 bp, although this is larger than most sites so far characterized). The human repressor must find its specific site within $(2.4 \times 10^9 \text{ bp})/(22 \text{ bp}) = 1.1 \times 10^8$ nonspecific sites. Thus the ratio of nonspecific to specific sites is $(1.1 \times 10^8)/(2.1 \times 10^5) = 519$ times greater in the human cell. Extrapolating from the *lac* repressor information, we estimate that $(519)(10 \text{ molecules}) = 5,190 \approx 5,000$ molecules of repressor will be needed per cell.

8.   *Positive Regulation*   A new RNA polymerase activity is discovered in crude extracts of cells derived from an exotic fungus. The RNA polymerase initiates transcription only from a single, highly specialized promoter. As the polymerase is purified, its activity is observed to decline. The purified enzyme is completely inactive unless crude extract is added to the reaction mixture. Suggest an explanation for these observations.

Answer   The catabolite-regulated operons are examples of positive regulation. Glucose is the preferred carbon source for many bacteria, and the operons for metabolism of other sugars, such as lactose and arabinose, are less active when glucose is available (even in the presence of inducer). The cAMP concentration is low when glucose is available. As glucose is depleted, [cAMP] increases and cAMP forms a complex with the catabolite activator protein (CAP). The cAMP-CAP complex binds to a specific site near the *lac* promoter and the *ara* promoter. In both cases cAMP-CAP increases the efficiency of transcription from the promoter under induced conditions.

At least three different explanations can be offered for the observed loss of activity of the new RNA polymerase upon purification. (1) Using the example of cAMP-CAP, perhaps the fungal RNA polymerase may be active on its promoter only in the presence of an activator protein. The efficiency of transcription initiation by the polymerase may be low for the purified polymerase, but high in the presence of the activator in the crude extract. (2) Some critical subunit of the RNA polymerase may be loosely associated with the rest of the polymerase and was dissociated during purification. (3) If the assay for polymerase activity during the purification scheme is *specific initiation* of transcription from the promoter, the decline in activity may reflect a loss of specific initiation but not a decline in nonspecific transcription. In this case, a specificity factor, similar to the σ subunit of the *E. coli* RNA polymerase, may have been lost at one step of the purification.

CHAPTER 28 Recombinant DNA Technology

1. **Cloning**

   (a) Draw the structure of the end of a linear DNA fragment that was produced by an *Eco*RI restriction digest (include those sequences remaining from the *Eco*RI recognition sequence).

   (b) Draw the structure resulting from the reaction of this end sequence with DNA polymerase I and the four deoxynucleoside triphosphates.

   (c) Draw the sequence produced at the junction if two ends with the structure derived in (b) are ligated.

   (d) Draw the structure produced if the structure derived in (a) is treated with a nuclease that degrades only single-stranded DNA.

   (e) Draw the sequence of the junction produced if an end with structure (b) is ligated to an end with structure (d).

   (f) Draw the structure of the end of a linear DNA fragment that was produced by a *Pvu*II restriction digest (as in (a)).

   (g) Draw the sequence of the junction produced if an end with structure (b) is ligated to an end with structure (f).

   (h) Suppose you can synthesize a short duplex DNA fragment with any sequence you desire. With such a synthetic fragment and the procedures described in (a) through (g), design a protocol that will remove an *Eco*RI restriction site from a DNA molecule and incorporate a new *Bam*HI restriction site at approximately the same location. (Hint: See Fig. 28-3.)

   (i) Design four different short synthetic DNA fragments that would permit ligation of structure (a) with a DNA fragment produced by a *Pst*I restriction digest. In one of these synthetic fragments, design the sequence so that the final junction contains the recognition sequences for both *Eco*RI and *Pst*I. In the second and third synthetic fragments, design the sequence so that the junction contains only the *Eco*RI or the *Pst*I recognition sequence, respectively. Design the sequence of the fourth fragment so that neither the *Eco*RI nor the *Pst*I sequence appears in the junction.

   Answer Type II restriction enzymes cleave double-stranded DNA within recognition sequences to create either blunt-ended DNA or sticky-ended fragments. Blunt-ended DNA fragments can be joined by the action of T4 DNA ligase. Sticky-ended DNA fragments can be joined by either *E. coli* or T4 DNA ligases, provided that the sticky ends are complementary. Sticky-ended DNA fragments without complementary sticky ends can be joined only after the ends are made blunt either by exonucleases or *E. coli* DNA polymerase I.

(a) The recognition sequence for *Eco*RI is (5′) GAATTC(3′), with the cleavage site between G and A (see Table 28-2). Thus, digestion of a DNA molecule with one *Eco*RI site

(5′)---GAATTC---(3′)
   ---CTTAAG---

would yield two fragments:

(5′)---G(3′)   and   (5′)AATTC---(3′)
   ---CTTAA   G---

(b) DNA polymerase I catalyzes the synthesis of DNA in the 5′ → 3′ direction in the presence of four deoxyribonucleoside triphosphates. Therefore, the ends of both fragments generated in (a) will be made blunt ended

(5′)---GAATT(3′)   and   (5′)AATTC---(3′)
   ---CTTAA   TTAAG---

(c) The two fragments generated in (b) can be ligated by T4 DNA ligase to form

(5′)---GAATTAATTC---(3′)
   ---CTTAATTAAG---

(d) The fragments shown in (a) have sticky ends with a 5′ protruding single-stranded region. Treatment of these DNA fragments with a single-strand-specific nuclease will yield DNA fragments with blunt ends

(5′)---G (3′)   and   (5′)C---(3′)
   ---C   G---

(e) The lefthand DNA fragment in (b) can be joined with the righthand DNA fragment in (d) to yield

(5′)---GAATTC---(3′)
   ---CTTAAG---

The same recombinant DNA molecule can be produced by joining the righthand DNA fragment in (b) with the lefthand DNA fragment in (d).

(f) The recognition sequence for *Pvu*II is

(5′)CAGCTG (3′), with the cleavage site lying between G and C (see Table 28-2).

Thus, a DNA molecule with a *Pvu*II site will yield two fragments when digested with *Pvu*II

(5′)---CAG(3′)   and   (5′)CTG---(3′)
   ---GTC   GAC---

(g) The lefthand DNA fragment in (b) can be joined with the righthand DNA fragment in (f) to yield

(5′)---GAATTCTG---(3′)
   ---CTTAAGAC---

The same DNA fragment can also be produced by joining the righthand DNA fragment in (b) with the lefthand DNA fragment in (f).

(5')---GAATTCTG---(3')
   ---GTCTTAAC---

(h) There are two different methods by which one can convert an *Eco*RI restriction site to a *Bam*HI restriction site on a DNA molecule.

Method 1: Digest DNA with *Eco*RI and then create blunt ends by using either DNA polymerase I to fill in the single-stranded region as in (b), or a single-strand specific nuclease to remove the single-stranded region as in (d). Ligate a synthetic linker that contains the recognition sequence of *Bam*HI,

(5')GCGGATCCCG (3')
  CGCCTAGGGC

between the two blunt-ended DNA fragments to yield, if the *Eco*RI digested DNA is treated as in (b)

(5')---GAATTGCGGATCCCGAATTC--- (3')
   ---CTTAACGCCTAGGGCTTAAG---

or if the *Eco*RI digested DNA is treated as in (d)

(5')---GGCGGATCCCG (3')
   ---CCGCCTAGGGC

Notice that the *Eco*RI site is not regenerated after the ligation of the linker.

Method 2: This method utilizes a conversion adaptor to introduce a *Bam*HI site into the DNA molecule. A synthetic oligonucleotide with the sequence (5')AATTGGATCC (3') is partially self-complementary, and it spontaneously forms the structure

 (5')AATTGGATCC
      CCTAGGTTAA.

The sticky ends of this adaptor are complementary to the sticky ends generated by *Eco*RI digestion so that this adaptor can be ligated between the two *Eco*RI fragments to form

(5')---GAATTGGATCCAATT---(3')
   ---CTTAACCTAGGTTAA---

Since ligation between DNA molecules with compatible sticky ends is more efficient than ligation between DNA molecules with blunt ends, Method 2 is preferred over Method 1.

(i) In order for the DNA fragments shown in (a) to be joined with a DNA fragment generated by *Pst*I digestion, a conversion adaptor has to be used; this adaptor should contain a single-stranded region complementary to the sticky end of an *Eco*RI-generated DNA fragment, and a single-stranded region complementary to the sticky end generated by *Pst*I digestion. The four adaptor sequences that fulfill this requirement are shown below, in order of discussion in the problem (N = any nucleotide):

(5')AATTCNNNNCTGCA
        GNNNNG
(5')AATTCNNNNGTGCA
        GNNNNC
(5')AATTGNNNNCTGCA
        CNNNNG
(5')AATTGNNNNGTGCA
        CNNNNC

Ligation of the first adaptor to the *Eco*RI-digested DNA molecule would yield

(5')---GAATTCNNNNCTGCA (3')
   ---CTTAAGNNNNG

This DNA molecule can now be ligated to a DNA fragment produced by a *Pst*I digest, which has the terminal sequence

     (5')G--- (3')
     ACGTC---

to yield

(5')---GAATTCNNNNCTGCAG---(3')
   ---CTTAAGNNNNGACGTC---

(Notice that both *Eco*RI and *Pst*I sites are retained.)

In a similar fashion, the other three adaptors can each be ligated to the *Eco*RI-digested DNA molecule, and the ligated DNA molecule can subsequently be joined to a DNA fragment produced by a *Pst*I digest. The final products with the use of the second, third, and fourth adaptor, are

(5')---GAATTCNNNNGTGCAG---(3')
   ---CTTAAGNNNNCACGTC---

(Notice that the *Eco*RI site is retained, but not the *Pst*I site.)

(5')---GAATTGNNNNCTGCAG---(3')
   ---CTTAACNNNNGACGTC---

(Notice that the *Pst*I site is retained, but not the *Eco*RI site.)

(5')---GAATTGNNNNGTGCAG---(3')
   ---CTTAACNNNNCACGTC---

(Notice that neither the *Eco*RI nor the *Pst*I site is retained.)

2. ***Selecting for Recombinant Plasmids***   When cloning a foreign DNA fragment into a plasmid it is often useful to insert the fragment at a site that interrupts a selectable marker (such as the tetracycline-resistance element of pBR322). The loss of function of the interrupted gene can be used to identify clones containing recombinant plasmids with foreign DNA. With a cosmid it is unnecessary to do this, yet one can easily distinguish cosmids that incorporate large foreign DNA fragments from those that do not. How are these recombinant cosmids identified?

> ***Answer*** λ phage DNA can be packaged into infectious phage particles only if it is between 40,000 and 50,000 base pairs in length. Since cosmid vectors are generally between 5,000 and 7,000 base pairs, they will not be packaged into phage particles unless they contain a sufficient length of inserted DNA.

3. ***DNA Cloning***   The plasmid cloning vector pBR322 (see Fig. 28-5) is cleaved with the restriction endonuclease *Eco*RI. An isolated DNA fragment from a eukaryotic genome (also produced by *Eco*RI cleavage) is added to the prepared vector and ligated. The mixture of ligated DNAs is then used to transform bacteria, and plasmid-containing bacteria are selected by growth in the presence of tetracycline. In addition to the desired recombinant plasmid, what other types of plasmids might be found among the transformed bacteria that are tetracycline resistant?

> ***Answer*** Ligation of the linear pBR322 to regenerate circular pBR322 is a unimolecular process and thus can occur more efficiently than the ligation of a foreign DNA fragment to the linear pBR322, which is a bimolecular process (assuming equimolar amounts of linear pBR322 and a foreign DNA fragment in the reaction mixture). The tetracycline-resistant bacteria would include recombinant plasmids and plasmids in which the original pBR322 was regenerated without insertion of a foreign DNA fragment; these would retain resistance to ampicillin. Also, two or more molecules of pBR322 might be ligated together with or without insertion of foreign DNA.

4. ***Expressing a Cloned Gene***   You have isolated a plant gene that encodes a protein in which you are interested. On the drawing below, indicate sequences or sites that you will need to get this gene transcribed, translated, and regulated in *E. coli*.

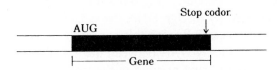

Answer The required sites and sequences are: an *E. coli* promoter, because *E. coli* RNA polymerase does not interact with eukaryotic promoters; a ribosome-binding site positioned at an appropriate distance upstream from the ATG codon, because eukaryotic mRNA does not utilize such a site for translation initiation; an operator site for regulation of transcription in *E. coli*.

5. *Identifying the Gene for a Protein with a Known Amino Acid Sequence* Design a DNA probe that would allow you to identify the gene for a protein with the following amino-terminal amino acid sequence. The probe should be 18 to 20 nucleotides long, a size that provides adequate specificity if there is sufficient homology between the probe and the gene.

$H_3{}^+$N-Ala-Pro-Met-Thr-Trp-Tyr-Cys-Met-Asp-Trp-Ile-
Ala-Gly-Gly-Pro-Trp-Phe-Arg-Lys-Asn-Thr-Lys---

Answer Recall from Chapter 26 that most amino acids are encoded by two or more codons. To minimize the degree of ambiguity in codon assignment for a given peptide sequence, we must select a region of the peptide that contains mostly amino acids specified by a small number of codons. Focus on the amino acids with the fewest codons: Met and Trp (see Fig. 26-7 and Table 26-4). The best possibility is the span of DNA from the codon for the first Trp residue to the first two nucleotides of the codon for Ile. The sequence of the probe would be

(5′)UGGUA(U/C)UG(U/C)AUGGA(U/C)UGGAU

The synthesis would be designed to incorporate either U or C where indicated, producing a mixture of eight 20-nucleotide probes that differ only at one or more of these positions.

6. *Cloning in Plants* The strategy outlined in Figure 28-17 employs *Agrobacterium* containing two separate plasmids. Suggest a reason why the sequences on the two plasmids are not combined on one plasmid.

Answer Simply for convenience; the 200,000 base pair Ti plasmid, even when the T DNA is removed, is too large to isolate in quantity and manipulate in vitro. It is also too large to reintroduce into a cell by standard transformation techniques. Single-plasmid systems in which the T DNA of a Ti plasmid has been replaced by foreign DNA (by means of low efficiency recombination in vivo) have been used successfully, but this approach is very laborious. The *vir* genes will facilitate the transfer of any DNA between the T DNA repeats, even if they are on a separate plasmid. The second plasmid in the two-plasmid system, because it requires only the T DNA repeats and a few sequences necessary for plasmid selection and propagation, is relatively small, easily isolated, and easily manipulated (foreign DNA is easily added and/or altered). It can be propagated in either *E. coli* or *Agrobacterium* and readily reintroduced into either bacterium.

7. *Cloning in Mammals* The retroviral vectors described in Figure 28-20 make it possible to integrate foreign DNA efficiently into a mammalian genome. Explain how these vectors, which lack genes for replication and viral packaging (*gag, pol, env*), are assembled into infectious viral particles. Suggest why it is important that these vectors lack the replication and packaging genes.

Answer The vectors must be introduced into a cell infected with a helper virus that can provide the necessary replication and packaging functions but cannot itself be packaged. The vectors packaged into infectious viral particles are used to introduce the recombinant DNA into a mammalian cell. Once this DNA is integrated into the chromosome of the target cell, the lack of recombination and packaging functions makes the integration very stable by preventing the deletion or replication of the integrated DNA.

SUPPLEMENT Protein Function as Illustrated by Oxygen-Binding Proteins

Problems

1. *Relationship between Affinity and Dissociation Constant* **(a)** Protein A has a binding site for ligand X with a K_d of 10^{-6} M. Protein B has a binding site for ligand X with a K_d of 10^{-9} M. Which protein has a higher affinity for ligand X? **(b)** Convert the K_d to K_a for both proteins.

2. *Efficiency of Oxygen Delivery* Carbon monoxide, a potent poison, binds with positive cooperativity to the heme of hemoglobin. The affinity of hemoglobin is about 200 times stronger for CO than for oxygen. Compare the efficiency of O_2 delivery in humans under the following two conditions. Under which condition is the efficiency of O_2 delivery lower? Why?

(a) A severely anemic person whose blood hemoglobin concentration is one-half the normal value.

(b) A person poisoned with carbon monoxide such that the number of O_2-binding sites is reduced to one-half the normal value (because every hemoglobin molecule has CO bound to two of its four hemes).

3. *Negative Cooperativity* Which of the following situations would produce a Hill plot with $n_H < 1.0$?

(a) The protein has multiple subunits, each with a single ligand-binding site. Binding of ligand to one site on the protein decreases the binding affinity of other sites for the ligand.

(b) The protein has a single subunit with two ligand-binding sites. Each of the ligand-binding sites has a different affinity for the ligand.

(c) The protein has a single subunit with a single ligand-binding site. As purified, the protein preparation is heterogeneous, containing some protein molecules that are partially denatured and thus have a lower binding affinity for the ligand.

4. *Affinity for Oxygen in Myoglobin and Hemoglobin* What is the effect of the following changes on the O_2 affinity of myoglobin and hemoglobin: **(a)** a drop in the pH of blood plasma from 7.4 to 7.2; **(b)** a decrease in the partial pressure of CO_2 in the lungs from 6 kPa (holding one's breath) to 2 kPa (normal); **(c)** an increase in the BPG level from 5 mM (normal altitudes) to 8 mM (high altitudes)?

5. *Cooperativity in Hemoglobin* Under appropriate conditions, hemoglobin dissociates into its four subunits. The isolated α subunit binds oxygen, but the O_2-saturation curve is hyperbolic rather than sigmoidal. In addition, the binding of oxygen to the isolated α subunit is not affected by the presence of H^+, CO_2, or BPG. What do these observations indicate about the source of the cooperativity in hemoglobin?

6. *Comparison of Fetal and Maternal Hemoglobin* Studies of oxygen transport in pregnant mammals have shown that the O_2-saturation curves of fetal and maternal blood are markedly different when measured under the same conditions. Fetal erythrocytes contain a structural variant of hemoglobin, hemoglobin F, consisting of two α and two γ subunits ($\alpha_2\gamma_2$), whereas maternal erythrocytes contain hemoglobin A ($\alpha_2\beta_2$).

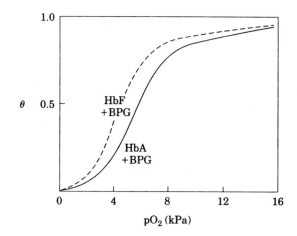

(a) Which hemoglobin has a higher affinity for oxygen under physiological conditions, hemoglobin A or hemoglobin F? Explain.

(b) What is the physiological significance of the different O_2 affinities? Explain.

(c) When all the BPG is carefully removed from samples of hemoglobin A and F, the measured O_2-saturation curves (and consequently the O_2 affinities) are displaced to the left. However, hemoglobin A now has a greater affinity for oxygen than does hemoglobin F. When BPG is reintroduced, the O_2-saturation curves return to normal, as shown in the figure. What is the effect of BPG on the O_2 affinity of hemoglobin? How can the above information be used to explain the different O_2 affinities of fetal and maternal hemoglobin?

7. *Effect on Oxygen Binding of a Mutation in Hemoglobin* A mutant hemoglobin is found in which the His residue at position H21 in the β subunits is replaced by Asp. Describe the effect of this mutation on the O_2-binding curve for hemoglobin. (Hint: See Fig. S–17.)

Solutions

1. (a) Protein B has a higher affinity for ligand X. The lower K_d indicates that protein B will be half-saturated with bound ligand X at a much lower concentration of X than will protein A. **(b)** Protein A has a K_a of 10^6 M^{-1}. Protein B has a K_a of 10^9 M^{-1}.

2. The binding of oxygen to hemoglobin shows positive cooperativity. Because both oxygen and carbon monoxide show the same cooperative behavior, the O_2-saturation curve for hemoglobin already bound to two CO molecules is described by the second half of the normal O_2-saturation curve—that is, it appears to be shifted to the left and to be less sigmoidal than the normal curve. This result has two important consequences. First, the number of available heme sites is reduced by half, and the capacity for O_2 delivery is also reduced by at least half. Second, oxygen binds much more tightly, and oxygen release at the tissues is therefore impaired. This dramatically lowers the efficiency of O_2 delivery and is generally fatal. The O_2-saturation curve of a severely anemic person, however, is unchanged. The hemoglobin molecules of an anemic person deliver oxygen with normal efficiency, but their lower numbers reduce the overall capacity for O_2 delivery. This condition is clearly less severe.

3. All three situations will produce a Hill coefficient, n_H, of less than 1.0. An $n_H < 1.0$ generally leads one to assume that situation (a) applies, which is the classic case of negative cooperativity. However, a closer examination of the properties of a protein exhibiting an apparent negative cooperativity in ligand binding often reveals situation (b) or (c), which represent common artifacts. Whenever two or more types of ligand-binding sites with different affinities for the ligand are present on the same or different proteins in the same solution, apparent negative cooperativity will be observed. In situation (b), the higher-affinity ligand-binding sites bind ligand first. As the ligand concentration is increased, binding to the lower-affinity sites will produce an $n_H < 1.0$, even though binding to the two ligand-binding sites is completely independent. Even more common is situation (c) in which there is heterogeneity in the protein preparation. This can be due to unsuspected proteolytic digestion by contaminating proteases, partial denaturation of the protein under certain solvent conditions, or many other causes. Because of the common occurrence of these artifacts, there are few well-documented cases of true negative cooperativity.

4. The affinity of hemoglobin for oxygen is regulated by the binding of the ligands H^+, CO_2, and BPG. The binding of each ligand shifts the O_2-saturation curve to the right—that is, the O_2 affinity of hemoglobin in the presence of the ligand is reduced. In contrast, the O_2 affinity of myoglobin is *not* affected by the presence of these ligands. Hence, the changing conditions only affect the binding of oxygen to hemoglobin: **(a)** decreases the affinity; **(b)** increases the affinity; **(c)** decreases

the affinity.

5. These observations indicate that the cooperative behavior—the sigmoidal O_2-binding curve and the positive cooperativity by ligands—of hemoglobin arises from interaction between subunits.

6. (a) The observation that hemoglobin A (maternal) is only 33% saturated when the pO_2 is 4 kPa, while hemoglobin F (fetal) is 58% saturated under the same physiological conditions, indicates that the O_2 affinity of hemoglobin F is higher than that of hemoglobin A. In other words, at identical O_2 concentrations, fetal hemoglobin binds more oxygen than does maternal hemoglobin. Thus fetal hemoglobin must bind oxygen more tightly (with higher affinity) than maternal hemoglobin under physiological conditions.

(b) The higher O_2 affinity of fetal hemoglobin assures that oxygen will flow from maternal blood to fetal blood in the placenta. For maximal oxygen transport, the oxygen pressure at which fetal blood approaches full saturation must be in the region where the O_2 affinity of maternal hemoglobin is low. This is indeed the case.

(c) The binding of BPG to hemoglobin reduces the affinity of hemoglobin for oxygen, as shown in the figure. The O_2-saturation curve for hemoglobin A in the absence of BPG is shifted far to the right when BPG binds (solid curves)—that is, the O_2 af-

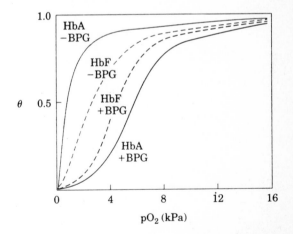

finity is dramatically lowered. The O_2-saturation curve for hemoglobin F in the absence of BPG is also shifted to the right when BPG binds (dashed curves), but not as far. The observation that the O_2-saturation curve of hemoglobin A displays a larger shift on BPG binding than does that of hemoglobin F suggests that hemoglobin A binds BPG more tightly than does hemoglobin F. It may be the differential binding of BPG to the two hemoglobins that determines the difference in their O_2 affinities.

7. The His residue at H21 in the β subunits is part of the BPG-binding site (see Fig. S–17). Substituting Asp for His at this position would decrease BPG binding and increase the overall affinity of hemoglobin for oxygen.